MATERIALS SCIENCE AND TECHNOLOGIES

RECENT ADVANCES IN NON-DESTRUCTIVE INSPECTION

Materials Science and Technologies

Additional books in this series can be found on Nova's website
under the Series tab.

Additional E-books in this series can be found on Nova's website
under the E-book tab.

MATERIALS SCIENCE AND TECHNOLOGIES

RECENT ADVANCES IN NON-DESTRUCTIVE INSPECTION

CAROSENA MEOLA
EDITOR

Nova Science Publishers, Inc.
New York

For permission to use material from this book please contact us:
Telephone 631-231-7269; Fax 631-231-8175
Web Site: http://www.novapublishers.com

LIBRARY OF CONGRESS CATALOGING-IN-PUBLICATION DATA

Available upon request.

ISBN: 978-1-61668-550-8

Published by Nova Science Publishers, Inc. † New York

CONTENTS

PREFACE

The ultimate desire of every manufacturer is to be able to see inside a finished product to ascertain its integrity without impairing it. At first sight, this may sound as *utopia*! Indeed, the application of some physical principles and modern technology permit to really travel inside the matter and see what is invisible to naked eyes.

And then, the word *non destructive* is becoming an ever more recurrent term in many fields from the *high-tech* development to the *past footsteps* conservation. In fact, it is necessary to assure fabrication of defects-free products and to check their in-service performance through planned maintenance, but it is also necessary to control the conservation state of artworks coming from the past. Words like non destructive inspection (NDI), non destructive testing (NDE), or non destructive evaluation (NDE) are often used with the same meaning. NDI should be understood as the way of determining material or part characteristics without disruption or impairment of serviceability. NDE and NDT may broadly be synonymous with NDI. The choice of one or the other is mainly a semantic point of view.

What is important is the final decision coming out! The part is free of defects, needs repair, or must be rejected. Certainly, the decision goes through inspection, testing and evaluation since:

- *Inspection*: means look at
- *Testing:* mainly describes the action
- *Evaluation:* indicates what it has been found

WHAT TO DISCOVER?

In general, the word defect is used to mean something wrong, or undesired in a material. In fact, it comes from the Latin word *defectus* which means deficiency, weakness, lack of something for completeness.

A material may be characterised by several types of defects which can be grouped into two main categories as *micro*- and *macro*-defects. Micro-defects are all the defects which affect the material crystalline structure and are also called crystalline defects. These may be also considered as false defects since they give the material specific properties and then, they are exploited for technological purposes.

Macro-defects are real non desired defects. Generally, they represent a discontinuity in the material and may develop during manufacturing, and/or in service. Basically, incorrect manufacturing processes may cause presence of defects such as voids, porosity, cracks, foreign inclusions, etc. In service, material degradation occurs as mainly due to fatigue cycles, impact damage, adverse environmental conditions which cause corrosion and overall decay. Defects may be, in general, classified as:

- open defects such as cracks and voids;
- gas filled volumes like bubbles;
- solid slag inclusions (backing film, dust, foreign material fragments);
- lack, or excess of adhesive in bonded joints and in adhesive based structures.

Of course, a complete classification may be performed owing to specific types of materials and structures and to the specific field under consideration (transport industry, architecture, environment, etc.).

Techniques and Procedures

The oldest and cheapest method to check for defects consists in using human senses, which means visual inspection, percussion and auscultation. Visual inspection is perhaps the method every manufacturer uses automatically, without thinking at it as a particular method to verify the artefact correctness. In other words, a look at the finished object permits to check shape, dimensions and surface conditions and to reject parts that are wholly out of order. Percussion is mainly used in architecture to individuate areas of detached plaster; the bond quality is estimated from the acoustical response to a knock. Of course, results depend on the operator skill. Thus, for a more reliable decision it is better to use non-subjective techniques.

Different types of methodologies are today available and yet, as technology evolves novel materials are created and new types of defects need detection, more sophisticated devices are developed, and/or the existing instruments are improved. Many physical parameters can be exploited for non destructive inspection and evaluation of materials. In fact, there exist techniques that are based on thermal-, acoustic- and electric- signal type, which may be used for detection of many types of defects, being shallow, or deep. Basically, no technique is able to detect every type of defects in every type of materials and thus, the most effective technique for the specific application must be chosen.

Generally, non-destructive evaluation can be performed by an energy input in the material and analysis of the output. More specifically, first a certain signal that might be electric current, or heat flux, or ultrasound wave, or X-rays, or something else is delivered to the material surface, then either the material response to such a stimulus, or the modifications experienced by the signal itself through the material are observed and analysed.

The propagation of a signal inside a material depends on its inherent mechanical, thermal, chemical, acoustical, electrical and optical properties. Thus, a material is classified as good electrical conductor if it is capable of easily carrying an electrical charge, or insulator when the flow of electrical current through it is negligible. Analogously, a good thermal conductor differs from a thermal insulator by how simply heat flows through it. Again, sound travels fast and easily through some materials while it is quite jaded by other materials. These and others material characteristics are exploited for non-destructive evaluation. In an homogeneous material, a given signal is in part reflected, in part absorbed and in part transmitted according to the material reflectivity, absorptivity and transmissivity for such a signal. An anomalous behaviour is a clear symptom of material deficiency.

It is important to know how a specific material characteristic modifies with the material overall degradation. The possibility to evaluate the variation of a material property through the variation of another property may be very useful and convenient because it allows the choice of the most adequate and effective technique and also minimizes the number of experimental tests with economic benefits.

The capability of a non-destructive evaluation technique to detect a discontinuity (defect) in a material depends on two main factors:

- The properties of the host material (thermal, optical, electric, etc.) and thus on the suitability of the evaluation technique.
- The resistance a discontinuity opposes to the propagation of the test energy-signal. More specifically, the *readable quantity*, related to the induced perturbation, must be large enough to not be confused with the instrument noise.

Owing to its inherent characteristics, a material can be inspected by using the most effective signal-based technique, or a combination of techniques. Basically, an inspection involves two main aspects: the material degradation in terms of loss of original characteristics (variation of density, hardness, elasticity, etc.) and the presence of defects both superficial (scratches, loss of optical properties, etc.) and internal (voids, cracks, etc.). Sometimes, a synergic integration of techniques is compulsory.

The attention of this book is focused on: 3D laser scanning, infrared thermography, and sound-, electric-, magnetic- based methodologies to span from surface observations down to subterranean remnants. Basically, two main areas are considered which are aerospace industry and cultural heritage. This choice, at first sight, may sound bizarre; but, materials within these two areas span from a wide range of thermal, electrical, chemical, physical, properties that allow assessing limits of each technique as well probable integration of different techniques. Besides, for both areas NDI is very relevant: in the aerospace industry mainly for safety purposes; in cultural heritage not to destroy what our ancestors left to us.

This book bears witness for expertise in different non destructive methodologies by some research Italian groups and also allows for virtual connection between these groups from south to north Italy.

THE BOOK CONTENT

The first chapter deals with the use of *3D laser scanning* applied to the modelling of cultural heritage and collects some of the work by the research group at the University of Padua. High resolution laser scanning methodology is analyzed, together with classical topographic measurements and digital photogrammetric techniques, for 3D surveys. The purpose of these techniques is to construct digital three dimensional models of objects. More specifically, the object is first laser scanned to create a point cloud from which the shape is extrapolated. This is helpful in different domains:

- in industrial engineering better known as reverse engineering and mainly exploited for design purposes;
- in civil engineering and in architecture for structural analysis and monitoring of small modifications for documentation and restoration of buildings;
- in territory survey for monitoring of volcanoes, glaciers and landslides areas, shoreline, etc.;
- in archaeological and cultural heritage fields for virtual reconstruction of buildings and work of art;
- in medicine for metric survey of prosthesis and changing evaluation pre- and post-surgical interventions.

Thus, this methodology may be regarded with a twofold meaning: reproduce metrically correct models of also complex geometry and control variation of geometry resulting from degradation of materials and structures. Of course, this technique may be incorporated inside NDE techniques, but it is restricted to the object surface.

Then, the successive chapters supply information through theoretical basics and practical applications for choosing the most adequate device and procedure to go more in depth inside the object and to check the material conditions.

In this context, the second chapter reports on applications of the *ultrasound* methodology at the Italian Aerospace Research Centre (CIRA). Indeed, ultrasound has been for years the leading technique for control of materials in the aerospace industry; and perhaps, it is still the most popular non-destructive technique. In fact, it can be also used in architecture and in civil engineering as the third chapter proves. In particular, chapter three joins expertise on the same subject coming from two different working entities which are the ENEA research Centre in south Italy and an University Department in north of Italy. Going back to the method principle, it is based on the propagation of high-frequency sound waves inside solid materials. A local variation of material characteristics (acoustical impedance) affects the energy transmission; thus, the amount of energy that arrives at the receiver probe gives information about the material characteristics (density, stiffness, porosity). Its main disadvantage is being time consuming for which it is now competing with emergent faster methodologies. However, it remains the most effective technique for thickness measurements.

A faster technique is certainly *infrared thermography* (IRT) that allows for generation of surface temperature maps starting from thermal energy radiated by an object in the infrared band of the electromagnetic spectrum. It is a truly 2D non contact, non invasive methodology which can be used for a vast variety of applications and also for non destructive evaluation of materials and structures. For example, IRT may be used for diagnosis (in medicine, architecture, maintenance), or for understanding of complex fluid dynamics phenomena (flow instability, flow separation and reattachment), or for material characterization and procedures assessment which can help improving design and fabrication of products. As technology evolves, infrared systems offer new opportunities for application; indeed, any process which is temperature-dependent may benefit from the use of an infrared radiometer. In this book, as for the other techniques, two chapters are devoted to applications of infrared thermography as non destructive technique. Chapter 4 by the Department of Aerospace Engineering at the

University of Naples supplies basic theory and examples of applications to NDE of materials which are mainly used in the industrial field. In particular, the use of both pulse and lock-in techniques is illustrated. The attention is particularly devoted to the capability of lock-in thermography to detect several types of defects such as delamination, slag inclusions, and impact damage in different types of materials mainly glass and carbon reinforced composites, sandwiches and fibre metal laminates. Applications of IRT to the cultural heritage field are illustrated in chapter 5. This chapter focuses attention on the thermophysical parameters to be taken into account and on the processing algorithms for correct interpretation of thermal maps and in turn for effective investigation of the decay of artworks. The methodological approaches are discussed while presenting some of the work performed at the CNR research centre of Padua.

At last, *magnetic and electric* methodologies are discussed in chapters 6 and 7. Indeed, this family includes many techniques such as magnetic particle, eddy current, geoelectrics, etc., which exploit electric and magnetic properties of materials studying their behaviour under magnetic and/or electric fields. Depending on the type of technique and on the measurement procedure, it is possible to detect shallow anomalies at depth of millimetres (e.g., with eddy current), or to discover ancient remnants at high deepness of 5-10 metres (e.g., with geoelectric, georadar). Again, the aim is to exploit electric/magnetic sensors in completely different application fields. And then, the attention of chapter 6 is mainly focused on the magnetic-based sensors for eddy current measuring devices; efforts are made to investigate the capability of such sensors to match the requirements of the aeronautical field. In particular, this chapter collects the expertise on the use of different types of sensors (i.e., flux-gate, GMR, HTc-SQUID) by researchers at the Coherentia section of the CNR-INFM in Naples. Electric/magnetic based techniques also belong to geophysical methodologies which are mainly used for underground exploration (mineral and oil, archaeology, etc.) and for mapping areas at natural risks (landslides, earthquakes, volcanic eruptions, etc.) for prevention and mitigation purposes. In chapter 7 some electrical geophysical methods are described and examples of application to architectural structures are shown. More specifically, the results shown, by the Department of Earth Science of the University of Naples Federico II, relate to micro-geophysics applied to the control of the state of conservation of buildings.

TOWARDS THE FUTURE

Taking into account that today does not exist a technique which is effective in detecting every type of defect in every type of material and/or structure, the efforts of the ND community are ever more focused on an integration of techniques, or *data fusion*. To this end, some attempts were already made in Naples for either data validation, or complimentary information.

On one side, both infrared thermography and SQUID based eddy currents were used for investigation of impact damage of carbon fibres reinforced polymers CFRP[1] and of dispersion of nanotubes[2] inside a polymeric matrix.

1 C. Meola, G.M. Carlomagno, M. Valentino, C. Bonavolontà, Non Destructive Evaluation of Impact Damage in CFRP with Infrared Thermography and SQUID, in Proc. 2nd Int. Conference on Buckling and Postbuckling

On the other side, infrared thermography has proved helpful in quickly detecting shallow anomalies in architectural structures, while the electric-type geophysical methods help to follow the in-depth propagation of such anomalies and to localize other deeper anomalies[3]. A combined use of the proposed techniques may be helpful for non-destructive evaluation of architectural structures. Infrared thermography, as a remote imaging system, can be used for quick evaluation of large surfaces; what is more, being a non-contact technique, it can be used also for the study of precious artworks such as mosaics, frescoes, paintings, etc. The ultrasonic method can be used to analyse the in-depth conditions, such as the thickness of the walls and the characteristics of the support under the plaster. The electric-type geophysical methods are helpful for high-depth evaluation. In addition, such methods can be used, in conjunction with infrared thermography, for material characterisation at low-depth. Thus, within a combined use of the three different methodologies, it seems possible to outline and characterise the endogenetic degradation sources from micro- to macro-scales[4].

Behaviour of Composite Laminated Shell Structures, Braunschweig (Germany) 3-5 September, 2008, Paper n.42 CDROOM.

2 C. Bonavolontà, M. Valentino, C. Meola, G.M. Carlomagno, R. Volponi, I.D. Rosca, Non Destructive Testing of carbon nanotube reinforced composite using HTS-SQUID and electromagnetic techniques, Superconductor Science and Technology, 22, 2009.

3 G.M. Carlomagno, R. Di Maio, C. Meola, N. Roberti Infrared thermography and geophysical techniques in cultural heritage conservation, QIRT Journal 2 (1), 5-24, 2005.

4 C. Meola, R. Di Maio, N. Roberti, G.M. Carlomagno Application of infrared thermography and geophysical methods to the architectural field Engineering Failure Analysis 12 (6), 875-892, 2005.

In: Recent Advances in Non-Destructive Inspection
Editor: Carosena Meola, pp. 1-31
ISBN 978-1-61668-550-8

Chapter 1

3D LASER SCANNING SURVEYS IN THE MODELLING OF CULTURAL HERITAGE

Massimo Fabris, Ginevra Boatto and Vladimiro Achilli
Laboratory of Survey and Geomatics – LRG
Department of Architectural, Town Planning and Survey – DAUR
University of Padua, Via Marzolo, 9 – 35131 Padova, Italy

ABSTRACT

3D high resolution survey is now performed mainly with photogrammetry and laser scanning methodologies providing very dense and accurate point clouds. Alignment of all acquired points in a defined reference system generates a metrically correct 3D model of the objects. Photographic information can be associated to the model too. Terrestrial applications have recently demonstrated the potentialities of the techniques, which are useful in different fields.

In this chapter 3D survey methodologies are explained, discussed and applied. These techniques have been used in different domains: in engineering field, as a geometrical support to structural analysis and in the monitoring of small variations by comparison of different surveys, co-registered in the same reference system. In architectural field, photogrammetry and laser scanning techniques are widely used for documentation and as support for restoration of buildings, while in territory survey are used for cartographic map production and in the monitoring of deformation areas (volcanoes, glaciers and landslides areas, shorelines) comparing the surveys at different time. In archaeology and in the study of cultural heritage, they allow for virtual reconstruction of buildings and objects. Moreover, in medicine, a metric survey helps quantitative evaluation of prosthesis and pre- and post- surgical interventions conditions by comparison of different models. In the sequel, some applications are described, showing different aspects concerning the 3D survey in the framework of cultural heritage. The "Cavallo ligneo" of the "Palazzo della Ragione" in Padua (a wooden horse from a middle age economical and political building), allows focusing on reconstruction of lacunas in the data generated by operative limitations. The "Arche Scaligere" in Verona, is used for discussing the integration between laser scanning and digital photogrammetric data, thus evaluating the accuracy of the methodologies. The work on the "San Prosdocimo" Chapel, which is located in "Santa Giustina" Basilica in Padua, is focused on different procedures for texture mapping of digital images on the 3D models distinguishing between point cloud models and surface models.

INTRODUCTION

In the last years, terrestrial digital photogrammetry and laser scanning systems have been used for the 3D surveys of objects in different research fields. These methodologies provide metrically correct 3D virtual models, which are useful in many applications. Generally, they have been applied in the framework of architectural, archaeological and cultural heritage fields for structural analysis, recover and preservation of buildings and objects, and for cognitive and cataloguing issues [1-3]. Cultural heritage objects are sensitive to natural events. Their extremely high value and the importance to grant their accessibility, make protection of cultural heritage necessary, especially by mitigating their sensitivity to the damages derived from extreme environmental events.

Thus, metrical morphological features of historical objects are the first step in a study aimed at their protection.

In this chapter some applications of high resolution 3D survey in the field of cultural heritage are introduced, analysed and discussed. Amongst them, the "Cavallo ligneo" of the "Palazzo della Ragione" in Padua, which is a massive wooden horse, the "Arche Scaligere" in Verona, and particularly the "Cansignorio Ark", which is a monumental grave of the lord of Verona, "San Prosdocimo" Chapel in the "Santa Giustina" Basilica in Padua, which is covered by important wall paintings.

1. THE METHODOLOGIES

1.1. Photogrammetry

The photogrammetric methodology defines shape, size and position of objects using images acquired from different points of view. In general, using the metric properties of an image, together with the coordinates of several control points, the geometry and the position of an object can be described in a defined reference system.

The main feature is extraction of 3D coordinates of a large number of points using at least two photographs of the same object. Thus, using suitable geometric relationships, together with the 2D information of the image, it is possible to obtain information on the third dimension [4-5]. Photogrammetry can be divided into two classes: aerial and terrestrial (or close-range), with a camera – object distance within 100-300 m. In the first case the images are acquired from an airplane, while, in the latter one, at the ground. The theoretical approach involved is the same for both categories, but the operative procedure, the scale and the instruments are different.

1.1.1. A Brief History of Photogrammetry

The photogrammetry allows for reconstruction of the metric characteristics of objects using photos. The first photograph ever was performed in 1827 by Joseph-Nicéphore Niépce that registered an image product through an optical system. In 1837 Louis Daguerre introduced at the Academy of Art and Science of France the first positive image on metal sheets. These inventions stimulated many researchers to extract metric features of objects from the images for military purposes. The French colonel Aimé Laussedat is indicated as the founder of the metric photography: he built the first photo-theodolite and, in 1864, carried out the first photogrammetric restitution of an architectural survey and the production of a topographic map of Paris, using photos acquired from roofs of buildings. The photogrammetric methodology was born in the second half of XIX century as "analogical photogrammetry": with two images of the same object acquired from different points of view it had been possible to geometrically estimate the 3D coordinates of each point located in the overlapping area. In the restitution of the survey, using optical-mechanical instruments, an analogy between the acquisition conditions and the photographs was performed: two subsequent images have to be 3D placed in the same relative position in which they were acquired.

The photographic film were positioned in the same relative position of the acquisitions and the operations were rather complex and requiring skilled workers. Thus, the generated stereoscopic model was scalar but representative of the real object.

The great diffusion of photogrammetry took place with the development of the aviation for the territory survey (aerial photogrammetry) while the second spread happened with the advent of Personal Computers and the birth of the "analytical photogrammetry". From the 1960's, the optical-mechanical instruments were computer-integrated, performing many functions in a easy way with respect to the procedure with analogical instruments. However, photographs were still acquired with analogical film [6]. The last transformation was carried out between the eighties and nineties with the "digital photogrammetry": in this case the whole photogrammetric operations were performed through a digital process. Images can be acquired by analogic or digital cameras; but, since digital photogrammetry processes images in numerical format, the available film must first be digitalized with conversion from continuous to discrete data set. At the same time the intensity of the signal is associated to a grey or colour scale assigned to each pixel. The introduction of Personal Computer, together with specific software, simplified the restitution process. Thus, a wider diffusion of the photogrammetric methodology could take place, with smaller costs.

However, the survey phase has to be performed by skilled workers [7].

1.1.2. Theoretical background

In order to reconstruct the shape of the objects, a single image is insufficient. The aim of photogrammetry is to perform measurements on the images, thus obtaining the survey of the object.

The object reference system (X,Y,Z), in the object space, and the image reference system (ξ,η), in the image space (an image is 2D) are defined. P is a generic point in the space object and P' its position in the image space; $O\left(X_0,Y_0,Z_0\right)$ the acquisition centre of the image and PP (ξ_0,η_0), Principal Point, the projection of the acquisition centre on the image plan; FC (Fiducial Centre) the origin of the image coordinate system (Figure 1).

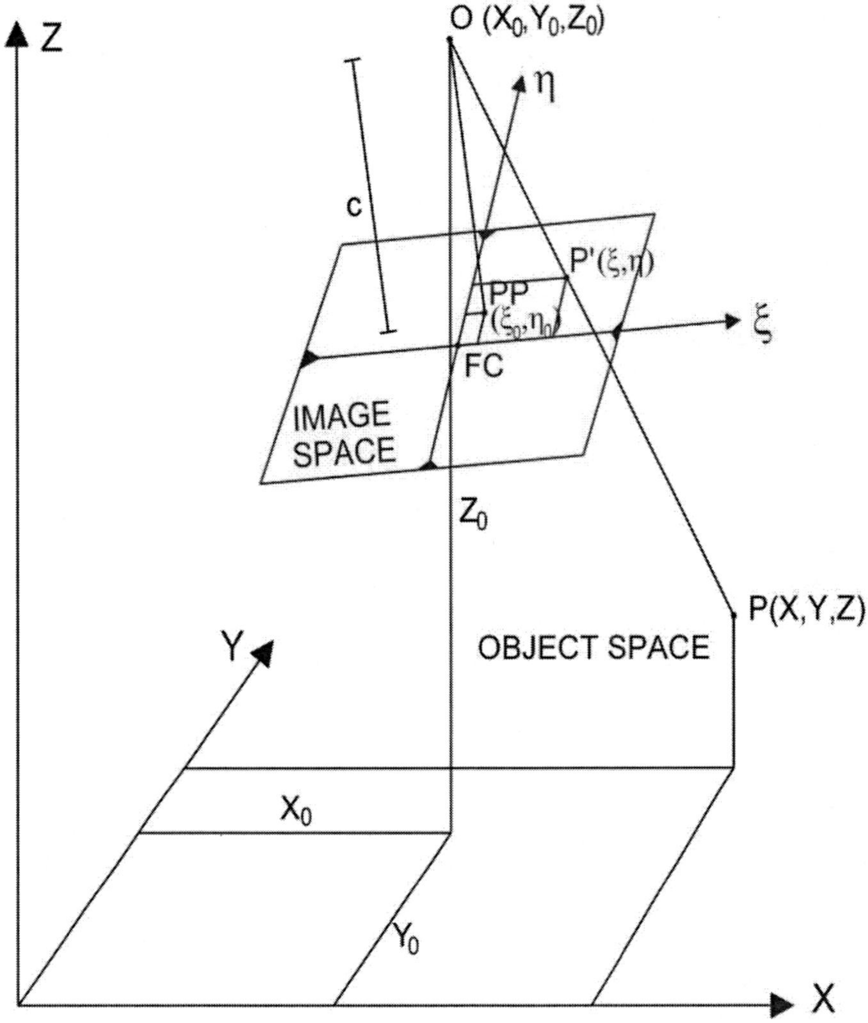

Figure 1. Scheme of the central perspective generation of an image.

With a single image, being the 2D position of the generic point P of the object (P') known, it is not possible to position the point in the object space: in fact, due to the fact that an image is a central perspective, the only information is provided by the half line going out from O and passing through P'. That is, the generic point P belongs to the half line but the position in the object space is unknown (Figure 1). If two images of the same object acquired from different points of view are available, geometrical ties, defining the 3D position of the point P in the object space, can be used (Figure 2). Thus, points P_1, P_2, P_3 and P_4 must lie over two half lines, for first and second images: the 3D intersection defines the position of the point in the object space. In this case, only with measurements performed on images it is possible to obtain the 3D survey of the object.

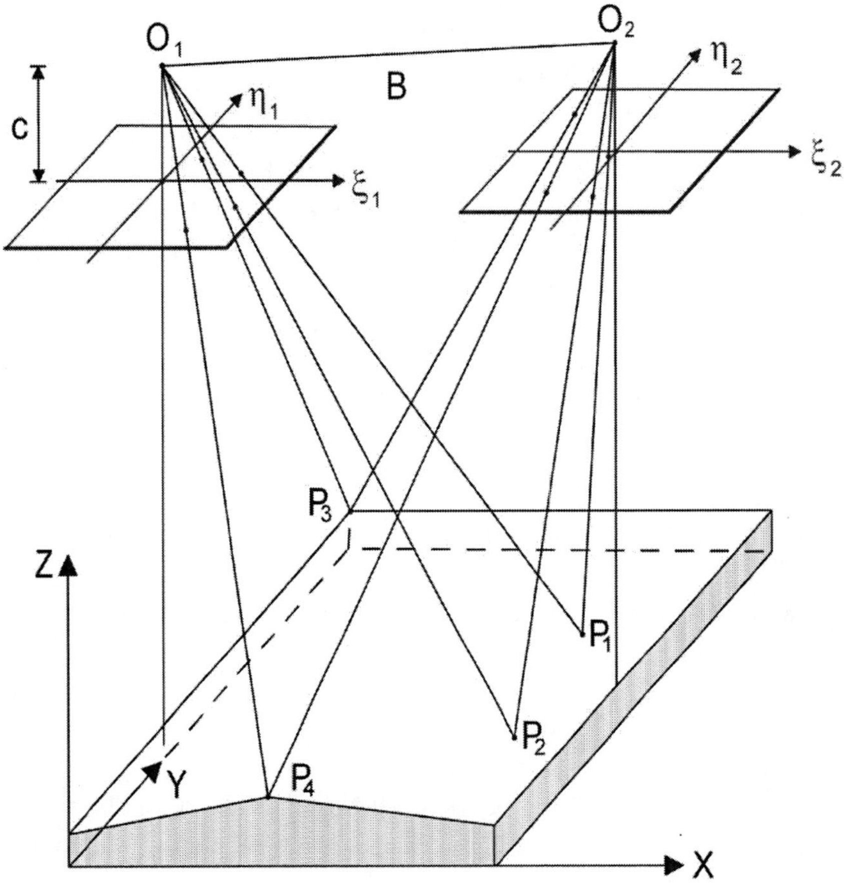

Figure 2. Scheme describing the photogrammetric measurements.

The relationships that connect the object reference system (X, Y, Z) with the image reference system (ξ, η) are:

$$X = X_0 + (Z - Z_0)\frac{r_{11}(\xi - \xi_0) + r_{12}(\eta - \eta_0) - r_{13}c}{r_{31}(\xi - \xi_0) + r_{32}(\eta - \eta_0) - r_{33}c} \quad (1)$$

$$Y = Y_0 + (Z - Z_0)\frac{r_{21}(\xi - \xi_0) + r_{22}(\eta - \eta_0) - r_{23}c}{r_{31}(\xi - \xi_0) + r_{32}(\eta - \eta_0) - r_{33}c} \quad (2)$$

where:

X, Y, Z are the coordinates of the generic point P in the object space;

X_0, Y_0, Z_0 are the coordinates of the acquisition centre of image in the object space;

ξ, η are the coordinates of the point P in the image space (P');

ξ_0, η_0 are the coordinates of the Principal Point (PP) in the image space;

c is the principal distance (focal length);

$r_{i,j}$ are the elements of the spatial rotation matrix between the two coordinate systems, depending on the three rotation angles (around the three coordinate axis) ω, φ, κ.

Analyzing Eqs. (1) and (2), which can be written for each unknown point P and for each image containing the same point, it is clear that, in order to determine the coordinates X, Y, Z, at least 3 equations have to be written (if other parameters are known). Thus, an image (that provides 2 equations) is insufficient: at least 2 images (4 equations) of the same point are needed.

The connection between the two coordinate systems is defined when the 15 parameters are determined; they are 6 coordinates of the acquisition centres of the two images (X_{01}, Y_{01}, Z_{01} for the first image and X_{02}, Y_{02}, Z_{02} for the second image), 6 rotation angles of the images ($\omega_1, \varphi_1, \kappa_1$ and $\omega_2, \varphi_2, \kappa_2$), the coordinates of the Principal Point (ξ_0, η_0) and the focal length (c): the last three parameters depend only on the camera used for the image acquisition. The parameters are the same for all of the images acquired with that camera, they change only if a different camera is used. ξ_0, η_0 and c (figure 1) depend on the metric characteristics inside the camera and are known as "inner orientation parameters". They are measurable values that are usually provided by the producer of the camera within a "camera calibration certificate" with high accuracy (the certificate provides also other parameters as lens radial distortion, the coordinates of the fiducial markers that define the FC point and ξ, η axis, etc.). Thus, if the camera is not calibrated, in order to reduce the unknown parameters, a preliminary calibration is necessary (in this case a "metric camera" is obtained). Moreover, calibration has to be performed periodically due to the little variations of the internal geometry.

The determination of the other 12 parameters (known as "external orientation parameters") requires that the user knows the coordinates of some points both in object and image reference systems (they are GCP – Ground Control Points). In (1) and (2) if $X, Y, Z, \xi, \eta, \xi_0, \eta_0, c$ are known, then $X_0, Y_0, Z_0, \omega, \varphi, \kappa$ are determined for the first and the second image. For each GCP, located in the area common to the two photographs, 4 equations can be written: 2 equations (1) (2) for the first image and 2 equations for the second image containing the same point. The 12 unknown parameters can be determined with a system of at least 12 equations: for this reason 3 GCP are required. If the coordinates of more GCP are known, a compensation can be performed that provides the 12 parameters and relative standard deviations.

The GCP have to be easily identifiable and measurable: they can be both natural (corner of buildings and sidewalks, crossroads, etc.) or artificial (using specific targets placed on the object surveyed). These points are measured with GPS (aerial photogrammetry) and/or classical topography (mainly for terrestrial photogrammetry) methodologies. For the complete survey of an object, each surface has to be photographed at least twice from different points of view in order to have each point visible on two images. The better accuracy is obtained with an overlap area of images of about 60-70%, thus, often, the complete coverage area requires many photographs acquired from different points: the progression of two or more images is called "strip" and the progression of two or more strips is named "block" [8-9].

During the elaboration phase, the stereoscopic model (that is the area in common between two adjacent images) is extracted, with two steps: inner orientation and external orientation. The internal orientation is performed in order to define the geometric characteristics inside the camera, using the camera calibration certificate to correct for distortion and to assign known coordinate values at specific points, such as fiducial markers and principal point. The external orientation is obtained, normally, by two subsequent steps: relative and absolute orientations (Figure 3). In the first case the aim is to create the stereoscopic model in a relative system using points common to both images (tie points). Usually, they are unknown natural points and, in this phase, the equations (1) and (2) are used for the transformation from the image reference system (ξ,η) to the relative reference system (x,y,z). Due to the unknown coordinates of the tie points, the relative reference system is arbitrary.

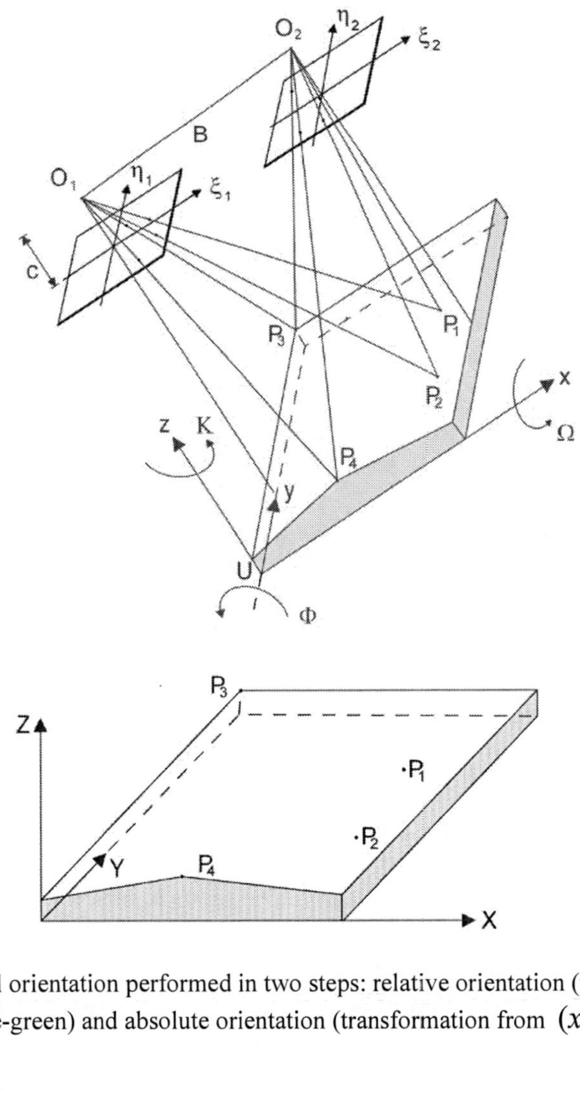

Figure 3. The external orientation performed in two steps: relative orientation (transformation from (ξ, η) to (x, y, z), blue-green) and absolute orientation (transformation from (x, y, z), to (X, Y, Z), green-black).

The absolute orientation is the transformation of the generated stereoscopic model into an external reference system defined by the coordinates of some GCPs (points with known coordinates), recognized on the images (from (x, y, z) to (X, Y, Z)); an *s*-transformation provides the registration of the two systems:

$$\begin{pmatrix} X \\ Y \\ Z \end{pmatrix} = \begin{pmatrix} X_U \\ Y_U \\ X_U \end{pmatrix} + m \cdot R \cdot \begin{pmatrix} x \\ y \\ z \end{pmatrix} \tag{3}$$

where (X, Y, Z) are the coordinates of the generic point P in the object reference system; (X_U, Y_U, Z_U) are the coordinates of the origin of relative reference system; m is the scale factor between the two systems; R is the spatial rotation matrix functions of the three rotation angles Ω, Φ, K between the (x, y, z) and (X, Y, Z); (x, y, z) are the coordinates of the generic point P in the relative reference system.

In this case the transformation can be performed knowing the 7 parameters $(X_U, Y_U, Z_U, \Omega, \Phi, \mathrm{K}, m)$ using, for example, 2 GCPs known in (X, Y, Z) and 1 GCP known only in (Z). In fact, with 7 coordinates known, a system of 7 equations (3) in 7 unknown parameters can be written. Thus, with this method less than 3 points have to be already known. In the generated stereoscopic model any 3D measure can be performed.

Photogrammetry produces coordinates of points in a 3D reference system, topographic maps with planimetric detail and contour lines, rectified photographs and derived maps and also digital images. Some by-products, useful in more applications, are Digital Elevation Models (DEMs), describing the elevation of an area (with buildings and vegetation too), Digital Terrain Models (DTMs), related to the terrain morphology, and Digital Surface Models (DSMs) describing the shape of a generic surface, both for aerial and close-range photogrammetry. These products are made up of many 3D points arranged in regular grids with step defined for specific requirements. The precision of a DSM depends on the image quality (if brightness is proper, if smoke or fog are present, if slope or vegetation are readable, etc.), resolution at ground (pixel dimension), grid size and depends also on the relationship between the distance and base of acquisition.

1.1.3. Digital Photogrammetry

Digital photogrammetry use numerical images, thus procedures related to artificial intelligence can be applied for the measurement of homologous points between the two images (GCP, tie points, DSM generation, etc.) that, in analogical and, partially, in analytical photogrammetry, must be performed manually. The smallest element of a digital image is the pixel: each pixel is characterized by a defined position on the image and by a value representing the intensity of colour or grey levels contained (radiometry).

The processing of stereo digital images can be automated using image matching procedures, based on well defined techniques relative to shape or grey/colour intensity for the same zones [4]. Generally, matching algorithms are classified into three principal groups: the ABM (Area Based Matching), the FBM (Feature Based Matching) and the RM (Relational

Matching) [10-14]. In ABM small windows composed of grey values are used as matching primitives to perform comparisons over homologous zones of images until the best correspondence is reached: cross correlation or Least Square Matching (LSM) methods can be used. In FBM, features (points, lines, edge elements, small regions, or other elements) are extracted in each image individually prior to matching them: the search is performed in terms of coordinates relative to the same objects characterized by a set of attributes (length, size, curvature, average brightness for regions, etc.). The RM concerns global features that are composed of different local features: the relations can be geometric (angle, distance, etc.) or radiometric (grey level between two adjacent regions, etc.). Digital photogrammetry is able to acquire many 3D points for high spatial resolution DSM generation through semi-automatic procedures, overcoming the problems of processing time and costs of the analytical approach.

The automatic research of homologous points in the ABM method is based on a correlation algorithm in term of radiometric values of pixels between a theoretic reference model (that can be, in aerial photogrammetry, a corner of building, a target, a crossroads, etc.) and a portion of image. Normally, the research is performed on a limited area of the image not to find multiple solutions (the algorithm can find many "corner of buildings" over all the image). For example, researching the position of the "reference model" of Figure 4a on the "research area" of Figure 4b, a linear correlation can be calculated for the 6 different positions of the reference model (position 4, 5, 6, 7, 8, 9) compared to the research area (Figure 4b, portion of image). Radiometry of each pixel of real image can be noisy due to the bad connection between the pixel and the object searched. It is used the coefficient of linear correlation:

$$r = \frac{\sum\left(g_1 - \overline{g_1}\right)\cdot\left(g_2 - \overline{g_2}\right)}{\sqrt{\sum\left(g_1 - \overline{g_1}\right)^2 \cdot \sum\left(g_2 - \overline{g_2}\right)^2}} \tag{4}$$

where g_1 is the radiometric value of the analyzed pixel in the reference model; $\overline{g_1}$ is the average of the radiometric values of the pixels in the reference model; g_2 is the radiometric value of analyzed pixel of research area; $\overline{g_2}$ is the average of radiometric values of the pixels of research area.

The maximum correlation is obtained in position 8, corresponding to the point automatically extracted (Figure 5a). However, an improvement of the result can be obtained at sub-pixel level (at position 7,5 circa in Figure 5b) weighting the pixels, assigning high value to the central pixel (that corresponds to the point searching in the image portion) and low values to the peripheral pixels.

The accuracy of digital photogrammetry mainly depends on the camera-object distance, image quality and resolution (pixel size); the latter is related to the digital camera characteristics or to the scanner resolution when the 'analogic' one is used. Dealing with terrestrial applications, where the camera-object distances are short (from few to some hundred meters), the precision of extracted DSMs ranges from millimetric to few centimetres. In the case of aerial surveys, the image scale may range from 1:1000 to 1:50000 or more; the precisions decrease from few centimetres to several meters [5]. Obviously, the image correlation method adopted for the DSM automatic extraction plays a key role in the minimization of errors of the photogrammetric working process [15].

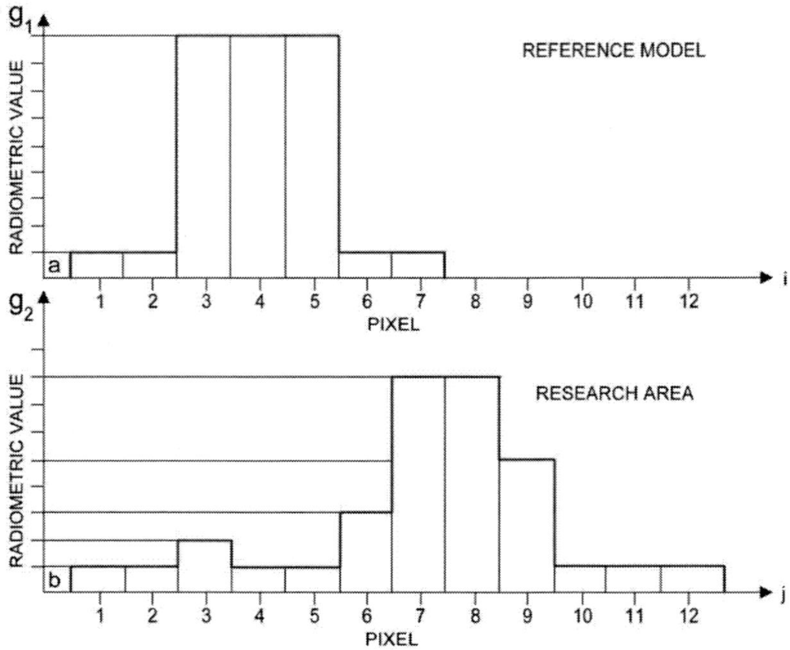

Figure 4. Linear correlation between a reference model (a) and a research area (portion of image)(b).

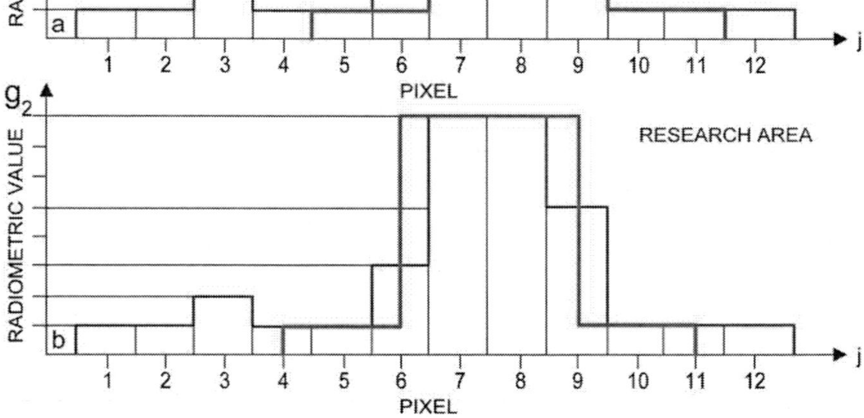

Figure 5. Improvement of linear correlation (a) working at sub-pixel level (b); pixels' weight decrease from the centre to the peripheries.

1.2. Terrestrial Laser Scanning

Laser scanning methodology can be considered an improvement of photogrammetry. It took place in the last decades: in fact, with the photogrammetric technique, the survey of a surface was performed during the elaboration phase on the acquired images measuring 3D points distributed on regular or irregular grid with defined size. In digital photogrammetry this procedure was carried out automatically by means of matching algorithms. The aim of laser scanning is to acquire 3D points directly during measurements *in situ*; for this reason is used the same principle for distance measurements of the total station in classical topographic survey. A laser beam generated by the instrument is shoot towards the object; after striking of the surface, a part of the beam is back reflected and is received by a device in the laser scanner: the distance is calculated by measuring the time required for the double distance between instrument and object (outgoing and returning wave) with the relationship:

$$D = v\frac{\Delta t}{2} \tag{5}$$

where D is the distance between the centre of the instrument and the point struck on the object, v is the velocity of the electromagnetic waves and Δt is the time measured. The same procedure is repeated for all points with different direction then, the surface is reconstructed with 3D points. In this case, the instrument uses Time Of Flight – TOF measuring principle: the time laser beam takes from the centre of laser scanner to the surface to be scanned together with its way back to the laser scanner is measured and the distance is calculated with Eq. (5). Being the two perpendicular small angles of the laser beam emission and the velocity of the electromagnetic waves known, the generic measured point can be positioned in the space using the polar coordinate system. Subsequently, the firmware of the instrument transforms these coordinates into the local cartesian reference frame with the origin coincident with the centre of the laser scanner.

Another approach is used by the triangulation laser scanner: the laser beam is projected with a certain angle to the surface of interest from the bottom one, while on top, the receiver of the back signal is placed. It is a Charge Coupled Device CCD, a photo-sensitive element receiving the reflection from surface. When distance between source, i.e. the rotating mirror, and CCD, direction of laser emission and back signal direction (emission and receiving angles) are known, then, by triangulation, the distance between laser scanner and measured point is calculated. Coordinates of each point are computed with reference to the coordinate system of the laser scanner, which is a three-dimensional right-handed cartesian system.

Laser scanning methodology provides large amount of data in a short time and with high accuracy, generally millimetric one or, for some specific instruments, also sub-millimetric. In terrestrial applications from a single station, laser scanners survey visible surfaces and, in order to reduce areas in the shade, acquisitions from different positions are necessary, also not reciprocally visible. In fact, the alignment of different scans is performed in the overlap areas between contiguous point clouds, or by means of integration with topographic classical methodologies. In the former case, before the 3D acquisitions, specific targets are placed on the surface to be surveyed and are measured both with laser scanning and topographic techniques in the same local reference system (planning and measuring of the local topographic network is necessary). The latter approach, the one using classic topography instruments, suggested for the alignment between scans without overlapping regions, for example, in the co-registration of point clouds concerning the external and internal portions of the building [16]. Accuracy of the terrestrial laser scanner mainly depends: on the device, deflecting laser beam through small rotations on perpendicular directions (normally formed by two mirrors in the configuration of optic galvanometer); on the measured range, that is the "Time Of Flight" (TOF) measure, or on the phase comparison between the outgoing and the back signal; on the resolution, that is the ability to detect small objects or object features in the point cloud (the combination between the smallest possible increment of the angle between two successive points and the size of the laser spot itself on the object); on the edge effects of the object: in fact, a variety of wrong points can be produced near the edges; on the reflectivity of the surfaces: normally, white surfaces will yield strong reflections whereas reflection is weak from black surfaces. The effects of coloured surfaces depend on the spectral characteristics of laser (green, red, near infrared). Shiny surfaces usually are not easy to acquire. Accuracy of terrestrial laser scanning depends finally on the environmental conditions (temperature, atmosphere, interfering radiation). These effects have been intensively investigated, studied and discussed for several laser scanner types [17-23].

The integration between terrestrial laser scanning and high resolution digital photogrammetric surveys, performed by means of metric, semi-metric or not metric cameras [5], is performed also with texture mapping procedures: in this case each single image is projected on the corresponding three-dimensional model portion; the outcoming product can be immediately and interactively used, due to the photorealistic information that can be georeferred and is metrically correct.

2. APPLICATIONS

2.1. Instruments

Several applications are presented. In the following, 3D surveys performed with terrestrial laser scanners Leica HDS (High Definition Surveying) 2500 and Konica Minolta Vivid 910 are obtained. The former has an accuracy of 4 mm in the distance measurements and 6 mm in positioning with a laser scanner – object range distances included between 1.5 and 50 m; the accuracy is 60 micro-radiant in angular measurement. The instrument acquires data within a 40° x 40° window (Field Of View – FOV); in this case the acquired pieces of information for a single scan are incomplete, but with different orientations of the scan-head from the same point, a broader view can be covered and areas in the shade reduced. The combination of small beam size with fully adjustable scanning density provides point to point spacing of 0.25 mm at 50 m: thus, the instrument can acquire smaller details, determine edge locations and accurately establish coordinates of targets, even within long distances. Targets, back-reflective at high reflectivity, are useful to align different scans acquired from different Points Of View (POV): after the first general scan, targets are identified and selected in the point clouds thanks to their different reflectivity respect to the surrounding materials. Subsequently, the firmware of the laser scanner performs detailed scans for each selected target (semi-automatic identification). Thus, the coordinates of the target are calculated as the mean of the detailed point clouds.

Konica Minolta laser scanner was used for the acquisition of the various details of the objects. It is a triangulation laser scanner and has two bull's eyes. The centre of the light-receiving lens of the digitizer is set as the origin of the coordinates. Accuracy of acquisition depends both on resolution of photo-sensitive element and distance between scanned object and laser scanner, two important factors determining quality of measures.

Topographic local network and laser scanning targets of each case-study were measured with Leica TC2003 total station, with accuracy of 1 mm + 1 ppm (1 mm for each km) for the distance and 0.15 mgon for angular measurements. These features make static and dynamic measuring procedures with high accuracy possible. At a distance of 100 m and under good visibility conditions, TC2003 can measure position of a point with an accuracy of under a millimetre.

Radiometric information of scanned surfaces were obtained with photographic surveys performed with Canon EOS 1 – Ds Mark II digital metric camera: the CMOS (Complementary Metal Oxide Semiconductor) sensor of 36 x 24 mm format provides images with 16.7 million pixels and gives TIFF file without compression until 50 MB with 24 bit depth colour.

2.2. The "Cavallo Ligneo" of the "Palazzo Della Ragione" (Padua)

2.2.1. Description

The "Cavallo ligneo" is a wooden horse statue, which is located inside the "Palazzo della Ragione" building in Padua, is large-sized: of height 5.75 m and of body circumference 6.20 m (figure 6). The object was designed so that a person could get inside it and sit comfortably: a trap-door is the access, while air and light get in from the mouth. In fact, the horse is hollow and built in order to resemble the hull of a ship, resting on the four sturdy legs which were made with a technique similar to segmented beams used today. The wooden boards are glued together on the long side to exploit the traction force of the wood in full, and minimise the warping typical of solid wood as much as possible. The horse was built in 1466 for a roundabout and commissioned by Conte Annibale Capodilista. It was transported in 1837 from Palazzo Capodilista to the Palazzo della Ragione, where it was dismantled and rebuilt almost entirely. Several parts, such as the head and the tail, were added afterwards.

2.2.2. The Laser Scanning Survey

The survey was performed integrating laser scanning and classical topography methodologies. Scans were acquired with Leica HDS 2500 laser scanner, from eight different stationing points (located in the four corners and in the medium points of the four walls) with 1 cm average grid size. Five laser scanning targets for each acquisition were used as control point. Coordinates were measured in a local reference frame by Leica TC2003 total station. Targets were placed outside the wooden horse but inside the window of acquisition of the laser scanner in order to acquire the whole object without artificial elements covering portions of it.

Figure 6. The wooden horse “Cavallo ligneo” of the “Palazzo della Ragione” in Padua.

2.2.3. Generation of 3D Point Clouds Model

The alignment of the point clouds generates 3D global model and was performed using Cyclone v. 5.4 software. This software uses three different approaches for co-registration of the scans: cloud registration, target registration and survey registration. In the former one, the point clouds are aligned using overlapping areas that must be large enough (at least 40%). Two scans are drawn up by choosing at least three homologous characteristics points between them: such points have to be easily recognizable on both point clouds. Subsequently, the ICP (Iterative Closest Points) algorithm application allows the refined adjust of the two scans. In this case, the co-registration of the point clouds is carried out by comparing the relative positions of homologous points. For each point recognized in the first scan, the algorithm searches for all the corresponding points closest to its normal in the second scan. Then it uses the two homologous points to estimate the new relative position between the two point clouds. The process is iterative and the convergence occurs when the distance between the scans is minor than a threshold value, depending on the resolution required to the survey, which, in this case, is 1 cm [24-27]. With target registration procedure, the alignment is carried out using common targets. Thus, the overlap between adjacent scans can be lower (about 30%) and the co-registration involves only the homologous targets. This procedure is more precise than the previous one which uses generic homologous points, because each target is acquired and measured by laser scanner with detailed point clouds at high resolution. In the latter method, i.e. survey registration, the alignment is performed using georeferred targets: the centre of each signal is measured in a local reference system by means of topographic methodologies. Subsequently, estimated coordinates are imposed to each target obtaining co-registration of point clouds in the same locale reference frame. In this case, no relevant overlapping percentage between adjacent scans is needed; nevertheless, minimum overlapping is required in order to guarantee continuity to the model (about 5÷10%). This method requires more acquiring time, but the final accuracy increases.

Survey registration method was used to align the scans of the “Cavallo ligneo”: each scan was roto-translated in a local reference system by imposing the coordinates of targets as control points. The global 3D model was processed with RapidForm2004 software (INUS Technology) merging data being overlapped from different adjacent scans.

2.2.4 Hole Filling

Some portions of the wooden horse statue were not acquired by the laser scanner because of the limited space available around the object (Figure 7). Several of these areas have no relevant dimensions, while others, for example the back and the top of the head of the horse, need reliable algorithms for filling these data holes. The reconstruction of these areas was conducted with RapidForm2004 software: the replenishing of lacunas can be performed with automatic, manual or combined procedures. In each of the three cases, the first step requires extraction of a polygonal mesh: the points of the 3D model are connected by simple small surfaces that, globally, reconstructs the shape of the object with a continuous and, generally, complex surface (that is the union of many plane faces).

The completely automatic procedure, which was applied on the back of the wooden horse, provided unsatisfactory results (Figure 8): the surface obtained is not fitting with the adjacent surfaces, several topological anomalies are visible too.

The semi-automatic procedure fills larger lacunas (Figure 9a), with respect of curvature of contiguous surfaces. At first, some faces in the hole have been reconstructed following the curvature of the surroundings. A scheme was thus built, dividing the lacuna in smaller portions and making hole filling easier (Figure 9b). Finally, automatic procedure was used (Figure 10). The closing of simple and small size holes was performed automatically, while for the complex and large lacunas semi-automatic procedures were used.

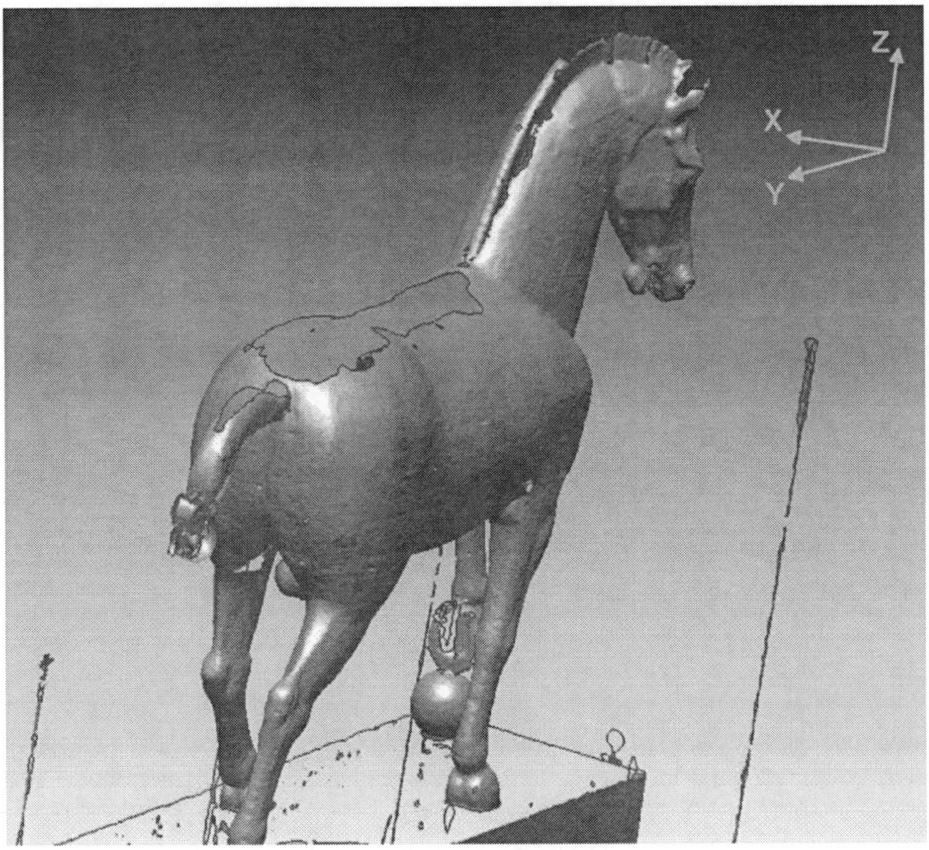

Figure 7. Global 3D surface model of the "Cavallo ligneo": mesh and portions with lacunas of data (red contoured areas) (after [3], modified).

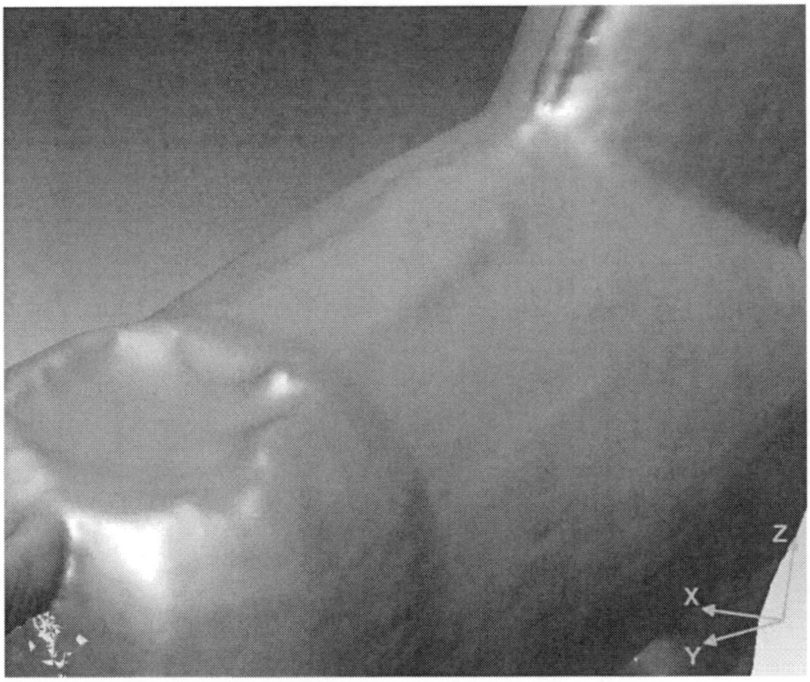

Figure 8. Filling lacunas with automatic procedure on the back of the wooden horse using RapidForm2004 software (after [3], modified).

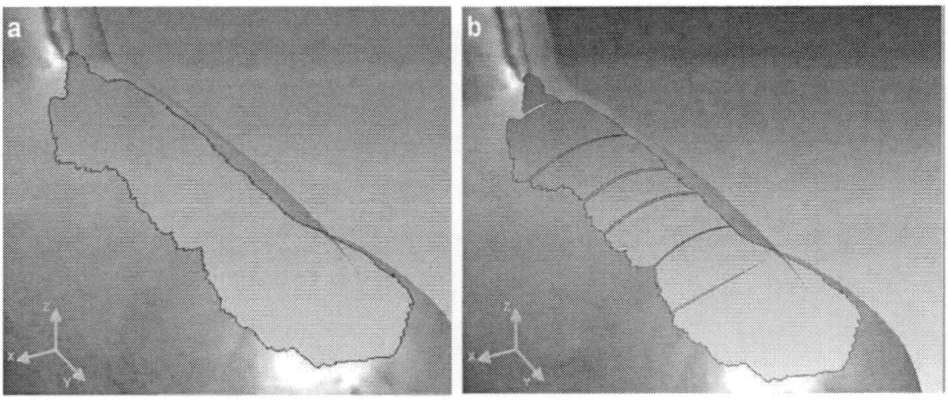

Figure 9. a) Lacunas on the back of the statue; b) Artificial subdivision in smaller lacunas using a transversal "skeleton" structure (after [3], modified).

2.2.5. 3D Surfaces Model Generation

Filling of lacunas has provided a global 3D surface model. The polygonal mesh was simplified reducing the number of faces but preserving the surface shape. In fact, flat areas do not need a large amount of data and thus help constructing a lighter model with low computational effort.

When adopting a 10% mesh-reduction, an increased smoothing is observed (Figure 11); moreover, the comparison between the decimated mesh and the original one provides residuals lower of HDS 2500 laser scanner accuracy.

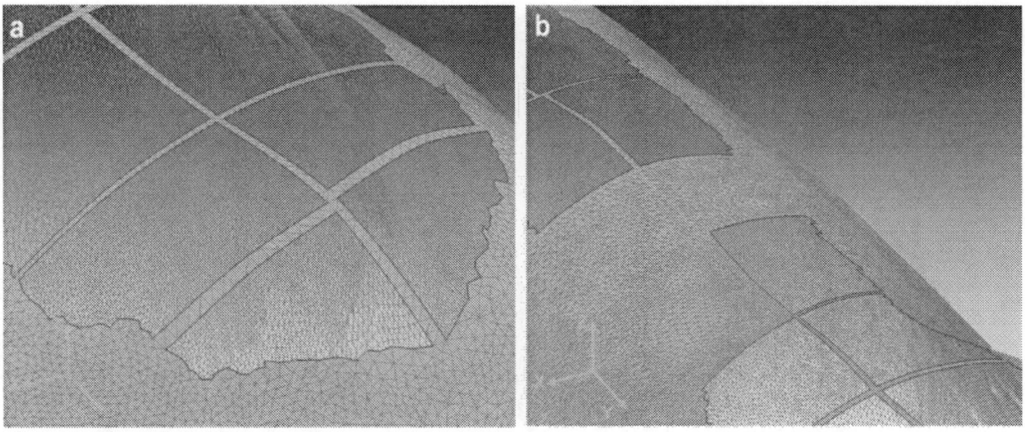

Figure 10. A) Generation of the longitudinal carrying skeleton; B) Automatic reconstruction of the new holes with small size (after [3], Modified).

Thus the mesh has been used to extract NURBS (Non-Uniform Rational Basis-Spline) surfaces. The operation is necessary because the mesh modification, if required for manual editing, is a slow and laborious procedure which may involve thousands of polynomial faces too. With NURBS surfaces, the procedure is easier when dealing with single elements covering the object: the final model is lighter too and easily manageable with other software.

The extraction of NURBS surfaces was performed using an automatic procedure. The number of control points and their distribution were decided afterwards (Figure 12) [3].

Figure 11. Final 3D reference model obtained by the survey integrated with semi-automatic procedures for the reconstruction of lacunas.

2.3. The "Arche Scaligere" – The Cansignorio Ark (Verona)

2.3.1. Description

The Cansignorio Ark is located inside "Scaligero" necropolis near Santa Maria Antica Church. There are nine tombs: the person which they are assigned to is known for the majority of them.

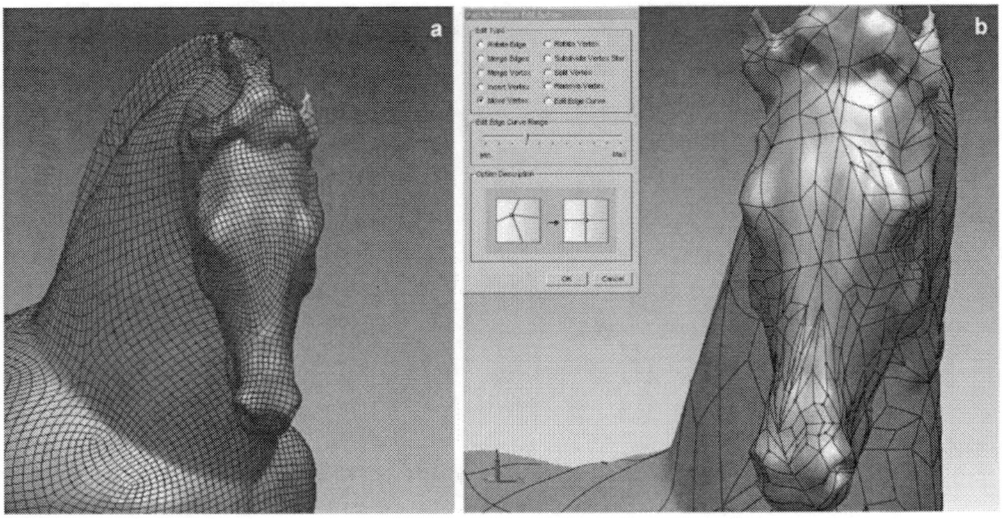

Figure 12. NURBS surfaces with control points management (after [3], modified).

Figure 13. a) Cansignorio Ark; b) Detail of the sarcophagus containing the corpse; c) Detail of the equestrian statue that completes the mausoleum (after [28], modified).

For the rest, and in particular for the graves on the ground, the attribution is more difficult due to the lack of engravings and epitaphs. The Arks were built between the XII and XIV century: the Cansignorio one, planned by the architect Bonino da Campione, was completed in 1376, one year after the purchaser died. The Ark, richly decorated, is placed in the northern-east angle of the necropolis: it is characterized by hexagonal plan with sides being about 3 m and height of about 15 m (Figure 13).

2.3.2. 3D Laser Scanning Survey and Data Elaboration

At first, a topographic reference network was placed around the monument and measured with Leica TC2003 total station in order to georeference laser scanning targets and photogrammetric signals in the same locale reference frame. Due to the limited available space around the object, the HDS 2500 laser scanner has required mobile platforms and acquisitions from windows at different floors in the overlooking buildings.

The stationing points were chosen to limit the shadow zones. Thus several scans were performed at 5.70 m, 8.20 m and 10.75 m of height; moreover, because of the morphological complexity of the object, the acquisitions were performed also inside the monument for surveyed niches, frames and every other element in shadow areas [28].

The scans were carried out with an average grid size of 0.5 cm, not exceeding the 50 m distance between laser scanner and object.

Before the acquisition, laser scanning back-reflective targets were placed on the Ark in order to perform a subsequent alignment using local reference system. In some scans, no targets were positioned due to the technical difficulty: in these cases a strategy that guaranteed a large overlapped areas between adjacent scans (about 60%) was used. In fact, subsequent alignment was performed using point clouds. Finally, 110 scans were performed acquired.

Co-registration was carried out with Cyclone v. 5.4 software: scans without targets were aligned with cloud registration procedure, while those with targets were aligned with survey registration method. The final 3D point cloud model of the Ark was obtained with maximum alignment error of 4 mm. Figure 14 shows ortho-projections of the model with respect to the planes of two opposite prospects, considering the contribution of the other prospects also.

Some details were acquired with the Konica Minolta Vivid 910 laser scanner in order to have a more specific virtual reconstruction of damaged portions. Figure 15 shows two examples related to the top of a pillar (a, b) and the bottom of a column (c, d).

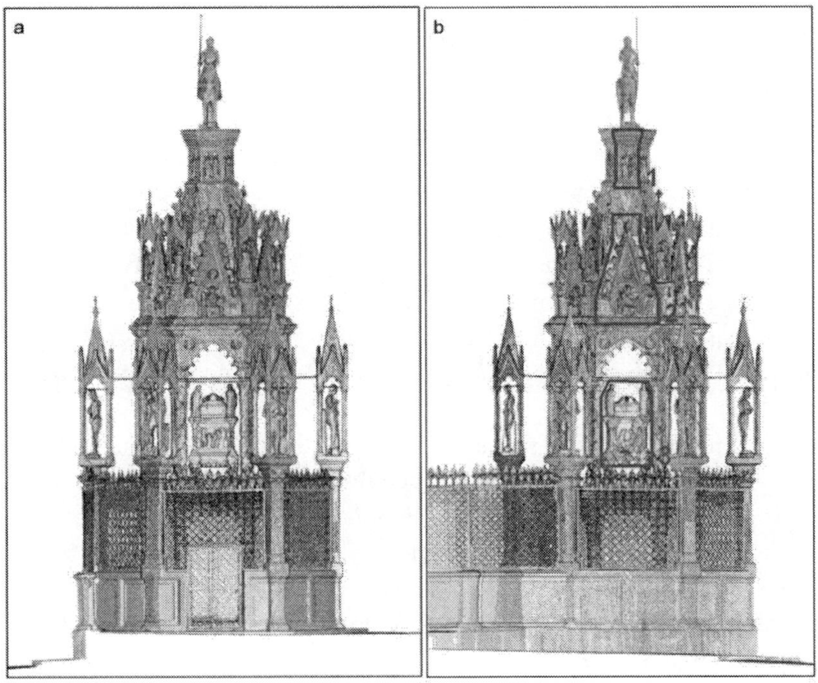

Figure 14. Ortho-projections of the points cloud global model of the Cansignorio Ark on two different opposite prospects (after [28], modified).

Figure 15. Virtual reconstruction of two details: a) top of a pillar with shading representation of the mesh; b) textured mesh of the same detail; c) bottom of a column with shading visualization; d) textured mesh of the same detail.

2.3.3. The Photogrammetric Survey

Terrestrial photogrammetric survey was carried out in order to generate high resolution orthophotos of the six prospects of the Ark. It was performed with Canon EOS 1 Ds Mark II metric camera and 53.23 mm calibrated focal length.

In the survey of the Cansignorio Ark, for each horizontal stripe, two images were taken with 70% overlap circa: with a mobile platform, for the vertical displacements, the other strips guaranteed an overlap in its vertical sequence of about 30% (this conditions required a minimum camera – Ark distance of 9 m and, consequently, pixel size on the monument of about 1 mm).

Due to the limited available space, in one case only the maximum camera – object distance was 7 m: the problem was solved by acquiring stripes formed by three images.

Natural points were chosen as control points, well recognizable on the images and uniformly distributed on the Ark. These points were measured with total station in the local reference frame defined by the topographic network.

The orientation and restitution of photogrammetric images were carried out with Socet Set (SoftCopy Exploitation Tool Set) v. 5.3 software. The automatic extraction of the DSMs on the stereoscopic models was performed with the adaptive ATE (Automatic Terrain Extraction) method and grid size of 0.5 cm on the areas concerning each prospect. In some portions a light smoothing was applied to the interpolated grid of the automatic DSMs with the aim of verifying the possibility for the manual editing reduction (Figure 16: areas concerning panels 1 and 2 of Figure 14).

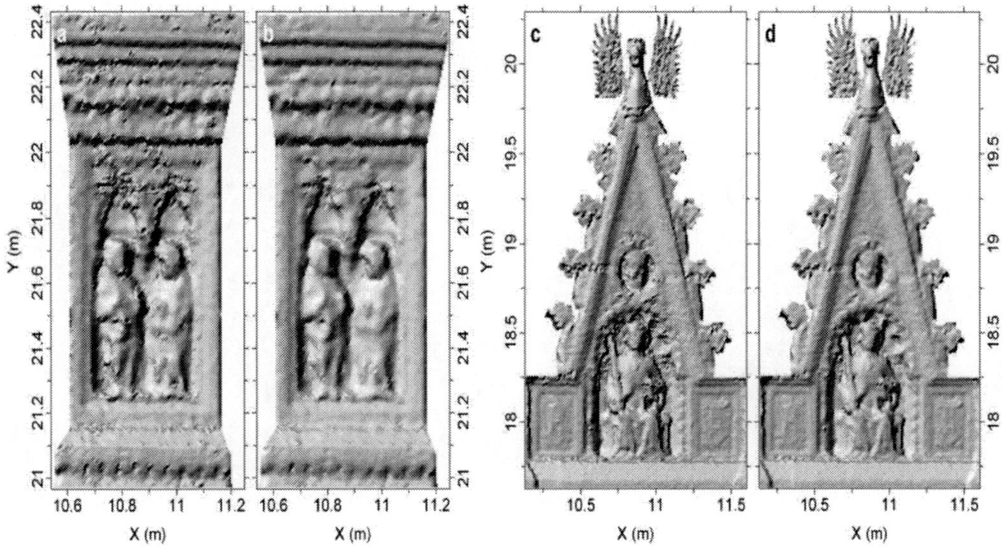

Figure 16. Digital Surface Models automatically extracted (a, c) and with a low smoothing (b, d) for the portions of panels 1 and 2 of figure 14 (after [28], modified).

The Figure 16 points out that in some areas the automatic extraction problems were reduced but without adaptation of the DSM to the real object morphology; moreover, in the battlements located in the lower portions of the figure, the smoothing application made the original result worse.

Thus, the DSMs automatically extracted were manually edited in the areas of unsatisfactory automatic correlation and in zones of uncorrected adaptation of the points to the monument morphology. Orthophotos with pixel size, on the Ark, of 2 mm were produced from the models obtained. In Figures 17 and 18, the final results of two prospects and an particular are shown.

2.3.4. Datasets Comparison

The application of laser scanning and digital photogrammetry terrestrial methodologies has provided two DSMs for each prospect. The two models were compared on three different areas indicated with 1, 2 and 3 in Figure 14. Before the direct comparison, the datasets were aligned in order to decrease co-registration residual error.

Orthophotos (Figure 19a, e, i), photogrammetric DSMs automatically extracted (grid size of 5 mm, Figure 19b, f, l), interpolated laser scanning DSMs (grid size of 5 mm, Figure 19c, g, m) and maps of the residuals distribution obtained by the comparisons (Figure 19d, h, n) are shown for each area.

Laser scanning models are characterized by more details with respect to the photogrammetric ones; however they appear more noisy probably because data have been obtained by the scans and also because of the interpolation algorithm used for the DSM generation (the kriging method was adopted for all the datasets).

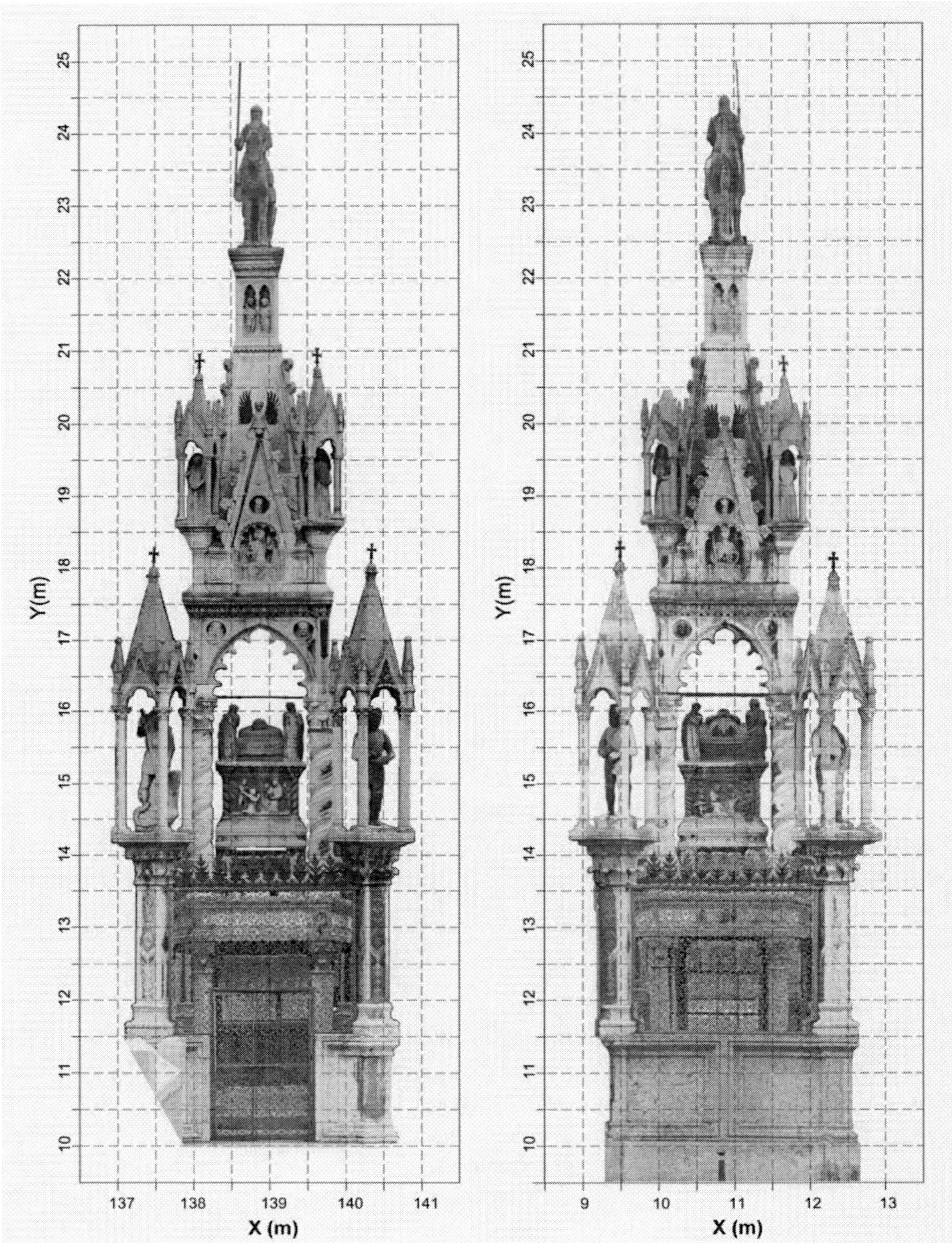

Figure 17. Orthophotos of the Cansignorio Ark concerning two different opposite prospects.

The comparison of models provided maps of residuals (quantitative differences between the two datasets) showing that the algorithm used in the correlation of the photogrammetric images is the least indicated when describing niches (concaves or convexes surfaces) and battlements of the analyzed portions.

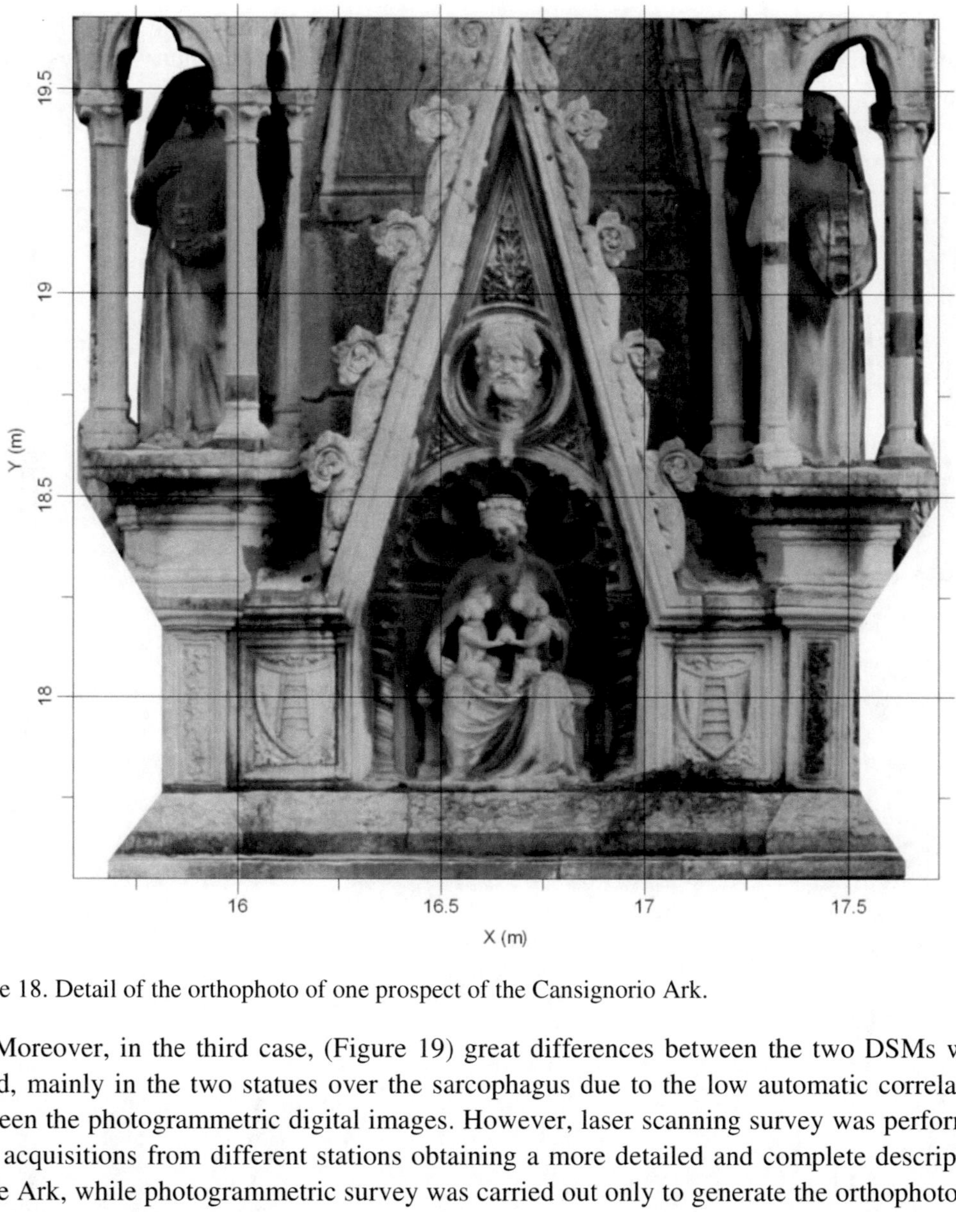

Figure 18. Detail of the orthophoto of one prospect of the Cansignorio Ark.

Moreover, in the third case, (Figure 19) great differences between the two DSMs were found, mainly in the two statues over the sarcophagus due to the low automatic correlation between the photogrammetric digital images. However, laser scanning survey was performed with acquisitions from different stations obtaining a more detailed and complete description of the Ark, while photogrammetric survey was carried out only to generate the orthophotos of the six prospects and not for detailed analysis.

2.4. "San Prosdocimo" Chapel ("Santa Giustina" Basilica)

2.4.1. Description

"San Prosdocimo" or "Santa Maria" Oratory, located inside "Santa Giustina" Basilica, is characterized by limited size: about 7x9 m of basis and maximum height in the top of the dome of about 10 m (Figure 20); it was built by Opilione, prefect of Teodorico, in the VI century B.C. and represents the most ancient architectonical building existing in Padua. Initially, it was decorated with mosaics narrating the life of San Martino bishop, but they collapsed because of an earthquake on 1117. Recently, the Chapel was restored after the second world war. The Chapel is an example of Byzantine art and San Prosdocimo's tomb, located in the oratory, shows typical "Ravennate Art" of the V – VI century B.C. (according to the Christian tradition, San Prosodocimo was the first bishop of Padua).

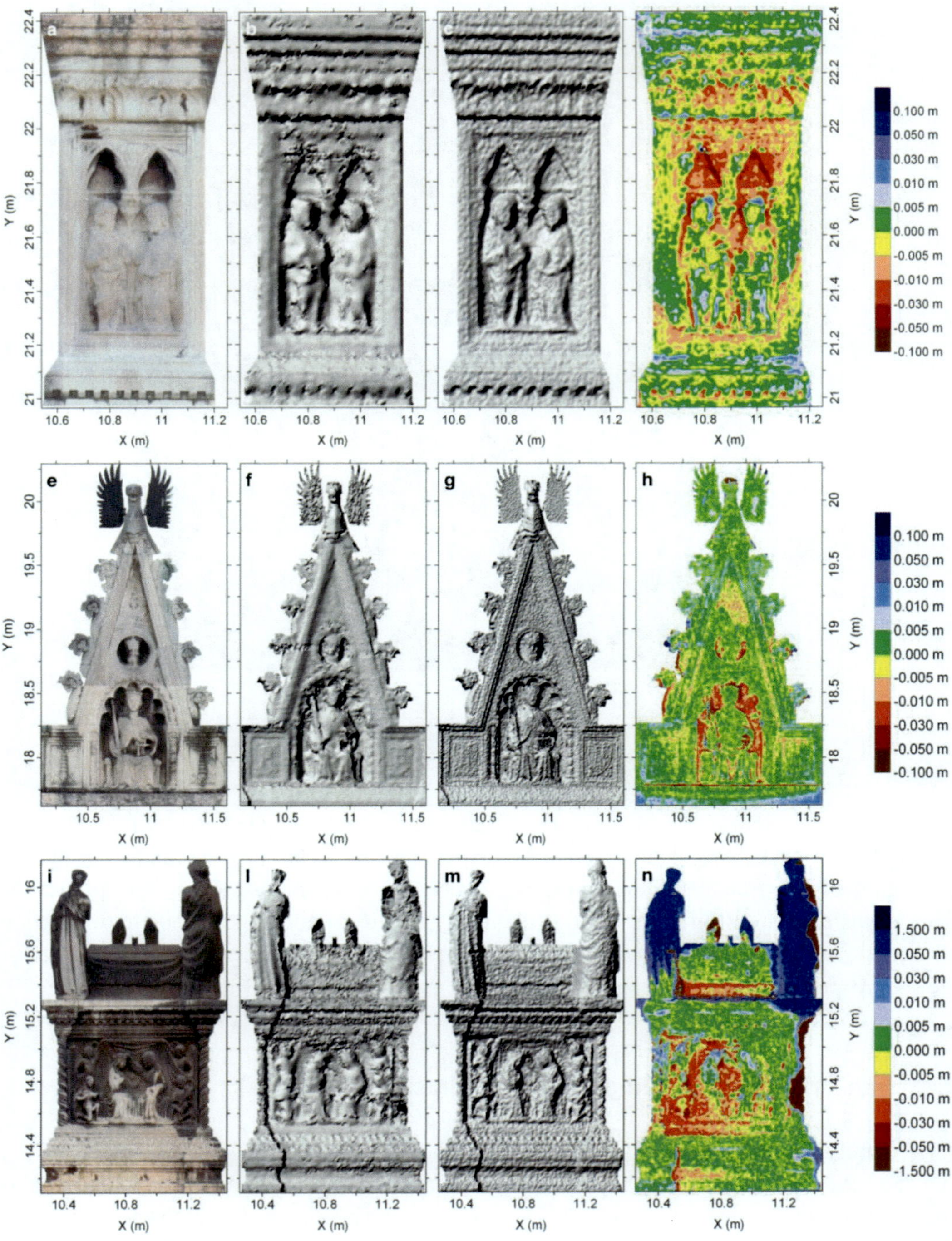

Figure 19. Orthophotos (a, e, i), automatic photogrammetric DSM (b, f, l), laser scanning DSM (c, g, m) and residuals map distribution (d, h, n) derived from the datasets comparisons (after [28], modified).

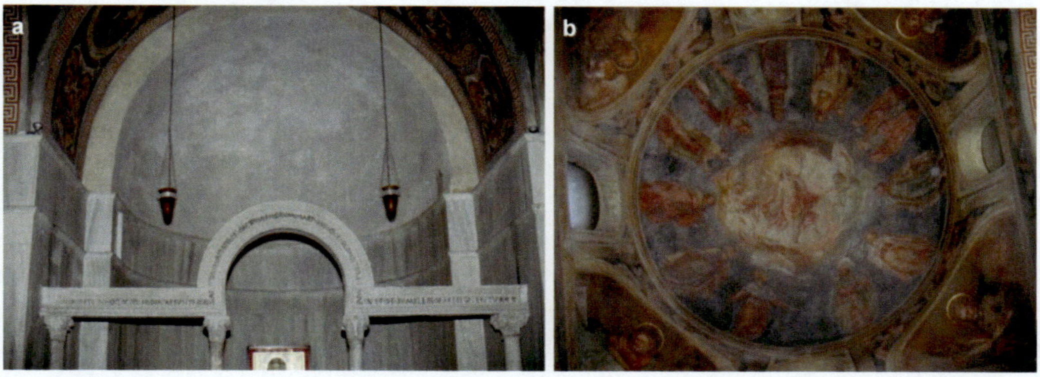

Figure 20. a) "San Prosdocimo" Chapel located inside "Santa Giustina" Basilica in Padua; b) The wall paintings on the dome of the Chapel.

2.4.2. The 3D survey

The laser scanning survey was carried out to obtain the 3D complete model of the inner part of the Chapel. A topographic network was planned and measured in a local system for georeferencing the laser scanning targets.

The scans were performed with an average grid size of 0.5 cm in the acquisition of 21 points clouds. The alignments were executed with Cyclone v. 5.3 software using cloud registration method for the co-registration of point clouds without targets laser scanning and the survey registration method for the alignment of the scans with georeferred targets.

Finally, 3D point cloud model of the Chapel was generated in the defined local reference frame with maximum alignment error of 3 mm (Figure 21).

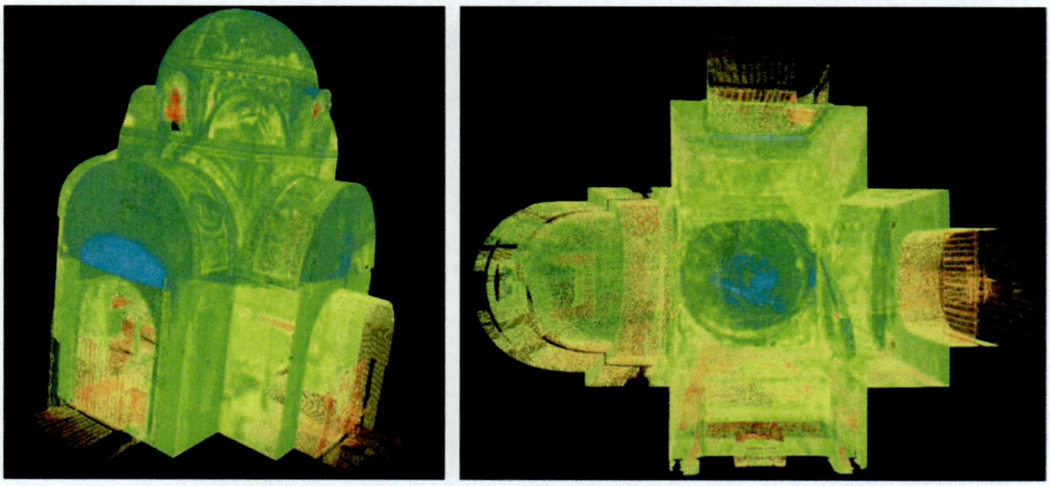

Figure 21. Final point cloud model of "San Prosdocimo" Chapel obtained by alignment of the single scans in the same local reference frame. For each point, the colour represents the reflectivity of the surface struck by the laser beam.

Each scan is made up of many points with three-dimensional coordinates; the laser scanner provides another parameter too, which is reflectivity of the surfaces, that is the response of each surface to the laser beam.

In fact, each material reflects the laser beam in a different way and the response depends on factors such as the incidence angle of the laser beam with the surface, the humidity of the surface, the colour of the surface, etc.

Generally, the reflectivity is represented with a colour: in Figure 21, some elements of the Chapel are recognized because characterized by different materials or different colours (Cyclone software). Using Reconstructor 2 software, points' reflectivity are viewed with greyscale (Figure 22).

The photogrammetric survey was carried out by keeping the metric camera at a maximum distance from the object of about 9 m; using a calibrated lens with of 53.23 mm focal length, a pixel's size on the object was about 1.2 mm.

The images were acquired also by the same station points of the laser scanning survey: in fact, the digital photographs are used to perform the texture mapping of the 3D laser scanning digital model obtaining photorealistic products useful in many applications.

2.4.3. Texture Mapping on 3D Model

The global point cloud was elaborated with Reconstructor 2 software: continuous surfaces, i.e. polygonal meshes, were extracted. They are formed by polygonal surfaces (generally triangles) that share one edge (and two vertexes) or one vertex only, with the vertex coincident to a scan point but more often only next to a scan point within a threshold value. Thus, the mesh surface represents an approximation of the point clouds derived from the survey. The result on the dome of the Chapel is shown in Figure 23. In this case reflectivity of the surfaces is lost.

Figure 22. Point cloud model of the dome of the Chapel with coloured points meaning reflectivity of the surface struck by laser beam (viewed with greyscale in Reconstructor 2 software).

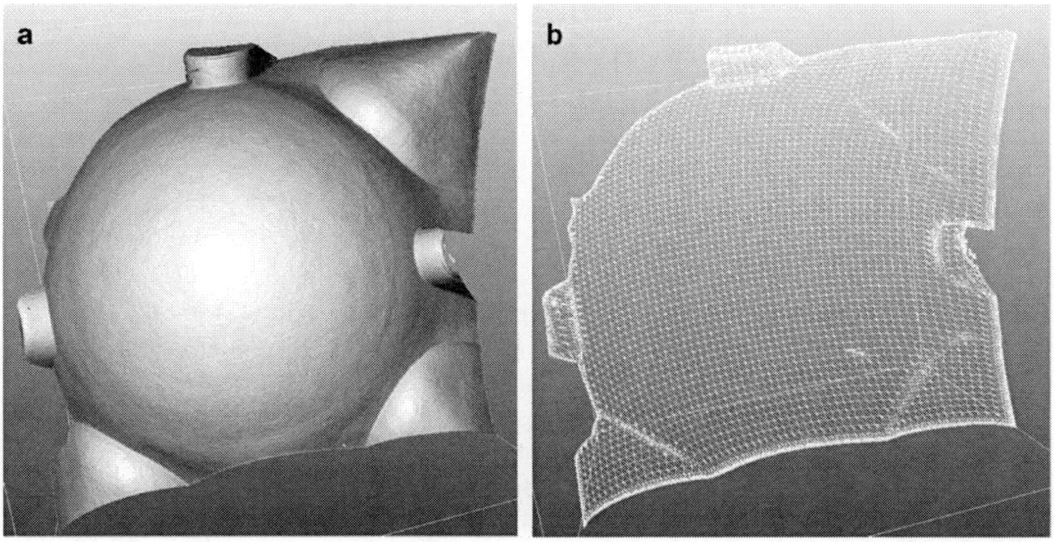

Figure 23. Different representation of the mesh surfaces: shading modality (a) and wireframe modality (b).

The 3D surface model was used to extract cross sections also, through the intersection between the 3D model with horizontal and vertical planes in the local reference system adopted (Figure 24).

Moreover, the same approach generates profiles, projecting the contours of the Chapel onto a reference plane. Cross sections and profiles reconstruct metric features of "San Prosdocimo" Chapel in AutoCAD software (Autodesk) which are very useful for the conservation and restoration of the structure.

The edge lines were automatically extracted comparing the normal lines of the surfaces and drawing the zones where the differences are major of a threshold (Figure 24).

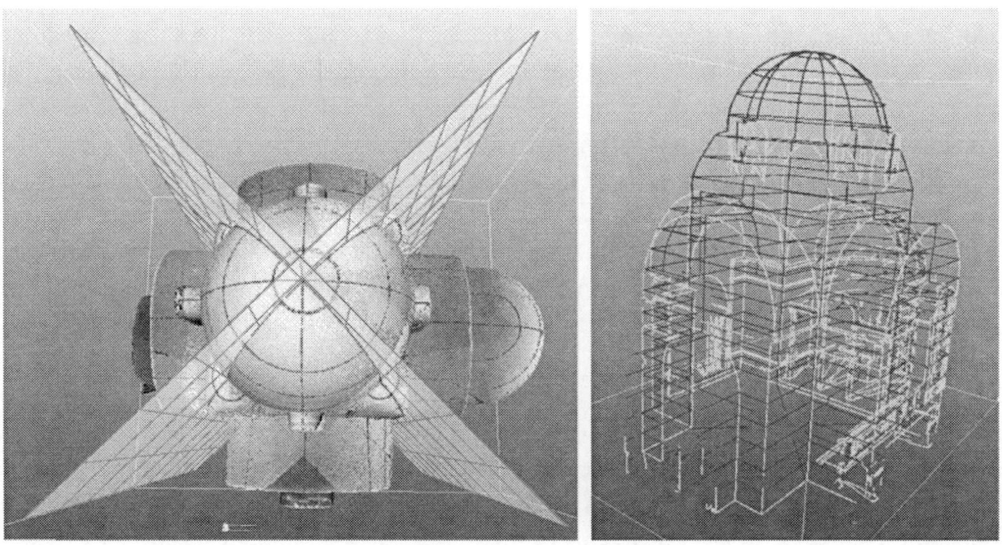

Figure 24. Extraction of the cross sections through the intersection of 3D surface model with vertical and horizontal plans (red lines). Edge lines are pointed out with yellow colour.

Texture mapping procedure can be done on the 3D global point cloud or on the 3D surface model: in the first case, for each point, the colour information derived from the digital image is associated. The homologous points, easily recognizable between the point clouds and the images, are identified and measured; the obtained product is useful only for very dense point clouds and, generally, global analysis, while for detailed study is unsuitable. In the second case, the pixels of acquired images are projected on the continuous surface allowing detailed analysis. However, the procedure for the homologous points measurements between the mesh surface and the image is characterized by more difficulty to find characteristic corresponding points (mainly for smooth, flat, concave or convex surfaces because the reflectivity information are loss after the surface generation). Moreover, the product is generally characterized by lower accuracy due to the approximation of the mesh. In "San Prosdocimo" Chapel case-study an integration of the two procedures was carried out to improve the final texture mapping on the mesh surface. At the beginning the texturing was performed on the global point cloud. Then, the images were calibrated directly on the 3D model using several homologous points and determining the centres of the projection for each image (Figure 25a). Subsequently, the calibrated images on the global points cloud were projected on the triangulated mesh improving the final result (Figure 25b). Thus, the disadvantage of the procedure for the texture mapping on the 3D surface model was solved. The figure 26 shows final texture on the dome of the "San Prosdocimo" Chapel (characterized by relevant wall painting) obtained by the integration of the two procedures.

3. Conclusion

In this chapter high resolution laser scanning methodology was analyzed, together with classical topographic measurements and digital photogrammetric techniques, for 3D surveys in the framework of cultural heritage. Three different applications were described and discussed in order to analyze all of the procedures that extracted precise and accurate 3D models with colour information derived by digital images texturing on the surface models.

The 3D survey of the "Cavallo ligneo" of the "Palazzo della Ragione" in Padua, gave the chance to test automatic and semi-automatic approaches for filling lacunas of data: in fact, the alignment of the acquired scans, performed with Cyclone software, provided a 3D model of the wooden horse statue with a large lacunas on the back of the horse due to the limited available space around it.

Due to the absence of other information, the reconstruction of lacunas of data was carried out by working on the curvature continuity with the surrounding surfaces. The better result was obtained with RapidForm2004 software by adopting a semi-automatic procedure: at first, a "scheme" has been created by the operator following the curvature and subsequently the holes, of reduced size, have been filled automatically.

The survey of Cansignorio Ark was performed integrating photogrammetric and laser scanning methodologies: in the first case high resolution orthophotos of the six prospects were obtained, while in the latter a 3D model was carried out with a grid size of 0.5 cm. The application compares models extracted with digital photogrammetry and laser scanning methodologies on three different significant areas: the differences were about of the same order of magnitude of the internal accuracy of each technique.

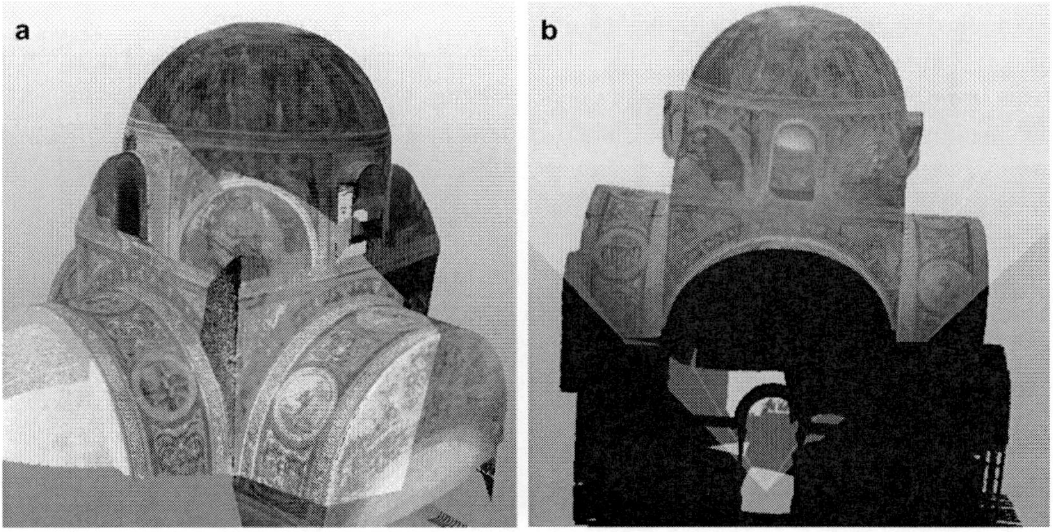

Figure 25 a) Texture mapping carried out on the global point cloud model; b) Projection of the images, previously calibrated, on the final 3D surface model.

Figure 26. The final texture mapping of the dome of the Chapel on the triangulated mesh.

In the last application, the texture mapping procedures used were analyzed: the 3D survey of the "San Prosdocimo" Chapel inside "Santa Giustina" Basilica was performed integrating laser scanning and classical topography methodologies. The 3D global surfaces model were obtained by aligning the single scans in a local reference frame, and extracting the triangular mesh. The texture mapping was carried out by calibrating the acquired images directly on the point cloud model; subsequently, the calibrated images were projected on the 3D surface model improving the final results.

The analyzed applications have demonstrated the potentialities of terrestrial photogrammetry and laser scanning methodologies for the accurate and high resolution 3D survey of objects in the framework of cultural heritage. Geometrical characteristics of an object are the first goal in every analysis aiming to its protection.

REFERENCES

[1] Bitelli, G. in Proc. of the 6ª National Conference *ASITA* (Perugia 2002), Perugia, Maggioli, *vol. 1*, pp IX-XXIV.

[2] Boehler, W.; Heinz, G.; Marbs, A. in Proc of the *XVIII CIPA* International Symposium Potsdam, CIPA Heritage Documentation 2001. http://cipa.icomos.org/fileadmin/papers/potsdam/2001-11-wb01.pdf.

[3] Fabris, M.; Achilli, V.; Bragagnolo, D.; Menin, A.; Salemi, G. in Proc. of the *XXI CIPA* International Symposium (Athens 2007), CIPA Heritage Documentation, 2007, pp 301-306 (Poster session).

[4] Selvini, A.; Guzzetti, F. *Fotogrammetria generale*. Edited by UTET, 2000, XII-516 pp. ISBN: 8802055319. ISBN-13: 9788802055312.

[5] Kraus, K. *Fotogrammetria (vol. 1)*, Ed. Levrotto and Bella, Torino, 1998.

[6] Bezoari, G.; Selvini, A. *Gli strumenti per la fotogrammetria. Storia e tecnica*. Edited by Liguori, 1999, 276 pp. ISBN: 8820726408. ISBN-13: 9788820726409.

[7] Crosilla, F.; Mussio, L. *Dall'analitico al digitale: nuovi sviluppi della fotogrammetria applicata all'ingegneria*. Edited by CISM, 1990, 242 pp. ISBN: 8885137040. ISBN-13: 9788885137042.

[8] Cundari, C. *Fotogrammetria architettonica*. Ed. Kappa, 1984.

[9] Bezoari, G.; Monti, C., Selvini, A. *La fotogrammetria per l'architettura*. Edited by Liguori, 1992, 332 pp. ISBN: 8820721651. ISBN-13: 9788820721657.

[10] Forstner, W. *Int. Arch. Photogramm. Remote Sensing*. 1986, *26,* 150-166.

[11] Shapiro, L.G.; Haralick, R.M. *Appl. Opt.* 1987, *26,* 1845-1851.

[12] Hannah, M.J. *Photogramm. Eng. Remote Sensing*. 1989, *55(12),* 1765-1770.

[13] Kraus, K. *Photogrammetry (vol. 2)*, Ed. Dummlers Verlag, Bonn, 1997.

[14] Wang, Y.. *Photogramm. Remote Sensing*. 1998, *53,* 154-165.

[15] Fabris, M.; Pesci, A. *Annals of Geophysics*, 2005, *48 (6),* 973-988.

[16] Fabris M.; Achilli, V.; Bragagnolo, D.; Menin, A.; Salemi, G. in Proc. of the *XXI CIPA* International Symposium (Athens 2007), CIPA Heritage Documentation, 2007, pp 307-310.

[17] Balzani, M.; Pellegrinelli, A.; Perfetti, N.; Uccelli F. in Proceedings of the *XVIII CIPA* International Symposium (Potsdam 2001), CIPA Heritage Documentation, pp 445-453.

[18] Boehler, W.; Bordas Vicent, M.; Marbs, A. in Proc. of the *XIX CIPA* Symposium (Antalya 2003), Antalya, CIPA Heritage Documentation, pp 696-702.

[19] Lichti, D.; Stewart, M. P.; Tsakiri, M.; Snow, A.J. *International Archives of Photogrammetry and Remote Sensing*, 2000, *XXXIII, Part B5*, 485-492.

[20] Lichti, D.; Gordon, J. in Proc. of *FIG Working Week 2004* Athens, 2004. http://www.fig.net/pub/athens/papers/wsa2/WSA2_6_Lichti_Gordon.pdf.

[21] Schulz, T.; Ingensand, H. In Proc. of *INGEO 2004 and FIG* Regional Central and Eastern European Conference on Engineering Surveying, Bratislava 2004. http://www.group.slac.stanford.edu/met/Align/Laser_Scanner/SchulzT_TS2_Bratislava_2004.pdf.

[22] Staiger, R. in Proc. of *Pharaohs to Geoinformatics FIG Working Week 2005 and GSDI-8* Cairo, 2005. http://www.fig.net/pub/cairo/abstracts/ts_38/ts38_05_staiger_abs.pdf

[23] Wasklewicz, T.; Staley, D.; Volker, H.; Whitley, D. 2004 http://geography

[24] Bergevin, R.; Soucy, M.; Gagnon, H.; Laurendeau, D. *IEEE Transactions on Pattern Analysis and Machine Intelligence*, 1996, *18,* 540-547.

[25] Besl, P.J.; McKay, N.D. *IEEE Transactions on Pattern Analysis and Machine Intelligence*, 1992, *14*, 239-256.

[26] Chen, Y.; Medioni, G. *Image and Vision Computing*. 1992, *10,* 145-155.

[27] Rusinkiewicz, S.; Levoy, M. in Proc. of the Third International Conference on *3D Digital Imaging and Modelling*, Quebec City, 2001, pp 145-152.

[28] Gaudini, G.; Achilli, V.; Bragagnolo, D.; Fabris, M.; Menin, A.; Ongarato, F.; Salemi, G. in Atti della 10ª Conferenza Nazionale *ASITA*, 2006, *vol. 2*, pp 1095-1102. ISBN: 88-900943-0-3 (Poster session).

In: Recent Advances in Non-Destructive Inspection
Editor: Carosena Meola, pp. 33-57
ISBN 978-1-61668-550-8

Chapter 2

ULTRASOUND TESTING FOR AERONAUTICAL APPLICATIONS

Cinzia Toscano
C.I.R.A. Scpa – via Maiorise s/n 81043 Capua (CE), Italy

ABSTRACT

Ultrasound Testing (UT) to carry out NDE, is basically concerned with the study of the response of a material to the propagation of a high frequency pressure waves train. Since the first applications of ultrasounds, their effectiveness in metallic materials examination have been noticed, in fact UT allows the detection of both the defects and its positioning in the whole volume of the inspected components. Moreover, thanks to the continuous technological development, UT offers wide possibilities for automatic processing which has made this methodology very attractive for industrials applications.

In particular, in the aeronautical industry, the replacement of metallic materials with composite, for the production of structural components, has emphasized the need to assess the integrity of the manufactured parts, in order to accomplish the severe quality design requirements and to carry out in-service health monitoring.

All these reasons have made UT one of the most widely used tool to inspect composite materials in the aeronautical industry.

INTRODUCTION

Since the first discovery of the existence of sound waves with frequencies above the upper limit of human audibility, the capability of ultrasounds to transmit informations has been noticed and several attempts to take advantage of this property have been carried out.

In the past century, ultrasounds have been studied and used mainly for wireless communications, in medical diagnostics and for examination, without induced damaging, of materials in the metallurgical industry. Examinations without damages of the materials structure is the principle Non Destructive Evaluation (NDE) is based on.

This chapter is focused on the ultrasound applications for NDE of aircraft components. Through a short historical reconstruction of the principle attempts and experiments carried out by the scientists as well as the results achieved, a brief view of the multiple applications of ultrasounds is provided. Then, some basic theoretical concepts and the nowadays most commonly applied techniques are reported, as well as, the typical methods used to display the presence of the defects in the examined structures.

The introduction and the consequent application of composite materials in the aeronautical industry is described, and, the consequent fundamental role played by UT for their inspection is highlighted. The most used composite materials manufacturing processes are briefly described in association with the typical defects involved, and the specific ultrasonic techniques suitable for their detection are shown. As one of the most common and subtle damaging cause for the composite structures, the problems related to the impacts occurrence and the detection of the induced defects are pointed out. The most used UT techniques are described in relation to the investigated defects as well as the most used testing setups for manual ad automated inspections. Finally, a brief view of recent developments and future trends is presented.

1. Ultrasound for Nondestructive Evaluation

1.1. Historical Hints

Ultrasounds are special acoustic waves with frequencies above the upper limit of human audibility (approximately 20 kHz).

At the end of the XVIII century, the speed of sound in air was already a measured parameter and the sound inability to propagate in vacuum had been demonstrated. The existence of sound waves with frequencies inaudible by human hears was pointed out in 1794 when the Italian biologist, Lazzaro Spallanzani, demonstrated that bats can fly in the darkness thanks to their ability to detect echoes coming from objects in the environment. Some years later, in 1826, the Swiss physicist Jean-Daniel Colladon measured the speed of sound in the water of the Geneva Lake thanks to his underwater bell. In the following years, physicists were working to mathematically describe sounds propagation and their interactions with media but only in 1877, the famous treatise "The Theory of Sound" by Lord Rayleigh was published. In this treatise, for the first time, sound waves were mathematically described opening the way for the future theoretical developments and successive applications.

The first very high frequency sound reproduction is due to the English scientist Francis Galton in 1876, when he realized its so called Galton whistle, but the real breakthrough in the study of high frequency sounds was obtained in 1880 when the *direct piezoelectric effect* was discovered by Pierre Curie and his brother, Paul Jacques Curie, in Paris (France). They succeded in measuring the surface charges (electricity production) appearing on specially prepared crystals (tourmaline, quartz, topaz, cane sugar and Rochelle salt among them), when they were subjected to mechanical stress. The property that crystals exhibiting the direct piezoelectric effect would also exhibit the *converse piezoelectric effect* (mechanical stress in response to an applied electric field) was, firstly, mathematically deduced from fundamental thermodynamic principles by Lippmann in 1881 and, subsequently, the Curies brothers confirmed experimentally their theory. Thanks to this discovery, ultrasounds production and reception was then possible taking advantage of the piezoelectric crystals properties.

One of the first applications of the underwater ultrasound (*sonar*) detection was proposed in 1912 after the "Titanic" sinking. Within a month of the disaster, the English meteorologist Lewis Richardson [1] claimed in its patent the possibility to detect icebergs by ultrasounds, but the first working sonar system was realized in the United States by the Canadian Reginald Fessenden only two years later, in 1914. Fessenden was studying an apparatus to underwater transmit and receive electromagnetic signals, in order to allow communication between ships. He conceived and designed an oscillator made by a copper tube (a strong electromagnet) surrounding a central core made by a coil. To transmit low frequency ultrasonic waves, he joined the copper tube and a large diaphragm together. The alternate current flowing through the core coil induced a current in the copper tube, which started to move back and forth, causing the diaphragm vibration. If the moving diaphragm was put in contact with water, the mechanical waves propagation was obtained. The oscillator was also used as a receiver in fact, the incoming mechanical waves caused the diaphragm vibration, so, the copper tube magnet started to move back and forth. This induced a current flowing in the coil which was related to the received ultrasonic signal. This first sonar based on the Fessenden oscillator was able to detect an underwater iceberg 2 miles away.

In the same years, thanks to the application of diodes and triodes, ultrasonic apparatus with enhanced amplification were developed. This led the French physicist Paul Langévin and the Russian scientist Constantin Chilowsky (who was resident in France), to start developing devices based on ultrasounds to detect submarines [2] which were largely used during World War I. Such device, called *hydrophone* included an array of thin quartz crystals between two steel plates with a resonant frequency of almost 150 KHz.

Between 1929 and 1935, S.Y. Sokolov [3] demonstrated the reliability of metal components examination using methods based on ultrasonic waves, so he introduced the concept of Non Destructive Evaluation, NDE, i. e. materials examination without damaging them; this became the main application of ultrasounds during World War II. He proved that the ultrasonic signals, transmitted through metals, were affected by losses of energy; the ultrasonic energy variations were associated with the presence of flaws. Therefore the detection of flaws was obtained by measurements of the ultrasonic energy attenuation. In Sokolov time, the resolution of this method was still quite poor, then he suggested the use of reflected signals instead of the transmitted ones, but the technological instrumentations were still inadequate for an efficient experimental apparatus.

A technique for ultrasonic data visualization was introduced in 1936 in Germany by Raimar Pohlman [4]. His system was based the transmission of acoustic waves which were transformed in images by acoustic lenses. This apparatus was widely used during World War II in order to assure quality controls of the ammunitions for the flak in Berlin.

During the 1940's ultrasounds for investigation of materials was improved in the United States by Floyd Firestone, who was the inventor of the Reflectoscope [5]. He modified a radar instrument and developed an amplified transmitter with short pulses. In the same period, Donald O. Sproule in Great Britain and Adolf Trost in Germany, although they had no knowledge of each other, started to apply a transmission technique based on separated transmitter and receiver probes. They put the two probes in contact on the opposite sides of a plate, on the same axis, via a mechanical device; continuously flowing water was used as coupling media between the probes and the investigated surfaces. This technique is now commonly used to inspect materials and it is known as *Through-Transmission* or *Transmission technique.*

In Germany, during 1949, the “Reflectoscope” invented by F. Firestone was investigated independently by Josef Krautkrämer in Cologne and Karl Deutsch in Wuppertal. Josef Krautkrämer and his brother Herbert, both physicists, were able to develop ultrasonic instruments in their company the KRAÜTKRAMER, while Karl Deutsch, a mechanical engineer, found a partner in Hans Werner Branscheid to set up the KARL DEUTSCH. These two companies producing ultrasound instruments for flaw detection, within only one year entered into competition which endured since few years ago.

Even if this chapter is devoted exclusively to ultrasonic applications for NDT, it has to be pointed out that ultrasound applications for medical purposes have been developed in parallel with those devoted to the inspection of metal components. In fact, between 1920 and 1930 the use of ultrasounds for the ablation of tissues, such as carcinomas, was already been experimented by William Fry at the University of Illinois [6], and their application to rehabilitation were already been put in practice.

Around 1940, ultrasounds were taken into consideration as diagnostic tool. Between 1942 and 1947, Dussik presented some results [7] concerning human organs inspection. He put two transducers on the opposite side of the head of a patient. The inspected parts of the head with both transducers were immersed in water to effectively transmit ultrasonic energy. The received signals were impressed upon a photographic paper sensitive to heat variations.

Around 1949, doctor George D. Ludwig active as junior Lieutenant at the Naval Medical Research Institute in Bethesda, Maryland, achieved to measure the ultrasound speed in human tissues (almost 1540 m/s) which is still used today [8]. His studies provided an important contribution to built the scientific foundation of the medical application of ultrasounds. Some years later, in 1959, Professor I. Donald in Scotland succeeded to detect clear echoes from the fetal head, and within only few years, the pregnancy monitoring from beginning to end was possible. These were the first steps in human diagnostics which had led to what today is known as *echography* [9].

Since the introduction on the market of the first ultrasound instrument, it was evident that the primary purpose of ultrasound inspections for industrials applications was the detection of defects. Defect detection led to develop the "safe life" approach aimed to prevent and reject dangerous component from service in order to avoid disruptions.

During the 1970’s the technological developments allowed improving instrumentation performance leading to the detection of smaller defects. In the industrial production, this caused the increase in the amount of the rejected components although the probability of failure had not changed. Thanks to the development of the fracture mechanics theory has become possible to predict the growth rate of cracks under cyclic loading (fatigue). This has formed the basis for the new philosophy of "damage tolerant" design, i.e. the possibility to accept structures containing defects, if the characteristics of such defects were known and verifiable. Therefore, the detection of the defect type and size, became a tool to establish if the defects will not grow to a critical size which may produce failure. Quantitative details, about flaw size, was used as input for fracture mechanics predictions which provided information on the remaining life of components. This approach became particularly strong in the military and nuclear power industries leading the way to the emergence of Quantitative Non Destructive Evaluation (QNDE) [10].

1.2. Theoretical Principles

Sounds of frequency above the upper limit of human audibility are called Ultrasounds, US, i.e. mechanical waves which can propagate in media such as solids, gases and liquids. Sound waves can propagate in specific modes [11], which in solids are:

- Longitudinal mode: the particles motion is parallel to the propagation direction, i.e. the wave behaves as a compression and a consequent relaxation in the particles distribution as shown in Figure 1 a).
- Transversal mode: the particles motion is orthogonal to the propagation direction as shown in Figure 1 b).

A guided wave or a Rayleigh wave is generated if the particles motion is forced to remain on the surface of the solid (in Figure 1 c): in this case the particles move on an ellipse around the propagation direction. The wavelength is defined as the distance between two points in which the wave is in the same state, as, for example, the distance between two peaks in a sinusoidal wave. If the solid thickness is smaller or equal to the sound wavelength, the Rayleigh wave degenerates in Lamb waves (plate waves).

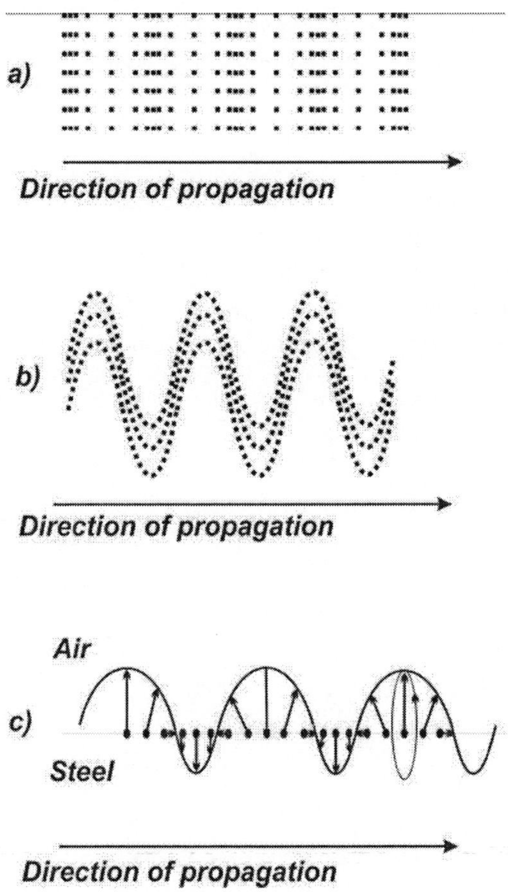

Figure 1. Particles motion in the Longitudinal wave mode (a), in the Transverse wave mode (b) and in the Rayleigh wave mode (c).

In gases and liquids only longitudinal wave can propagate. In particular human hears can perceive longitudinal waves with frequencies approximately between 10 Hz and 20 kHz.

Both longitudinal and transversal modes are characterized by their specific propagation velocity which depends on the physical properties of the materials. Longitudinal wave velocity is given by:

$$c_l = \sqrt{\frac{E}{\rho} \frac{1-\mu}{(1+\mu)(1-2\mu)}} \tag{1}$$

In Eq. (1), E is the Young modulus, ρ is the density and μ the Poisson's ratio; the transverse wave velocity is given by:

$$c_t = \sqrt{\frac{E}{\rho} \frac{1}{2(1+\mu)}} = \sqrt{\frac{G}{\rho}} \tag{2}$$

In Eq. (2), G is the Shear modulus.

Materials, from the acoustical point of view, can be divided in *sonically hard* or *soft* depending on their *acoustic impedance*, which is defined as:

$$J=\rho c \tag{3}$$

In Eq. (3), ρ is the density and c is the ultrasound wave velocity. For example steel, which has Z= 45×10^6 Ns/m^3, is sonically harder than water, for which Z=$1,5 \times 10^6$ Ns/m^3.

At the interface between two media A and B characterized by different acoustical impedance, the ultrasound wave are reflected and transmitted (Figure 2). The angles of reflection and transmission are linked to the ultrasound velocity in both media by the Snell's law:

$$\frac{\sin \alpha_I}{\sin \alpha_T} = \frac{c_I}{c_T} \tag{4}$$

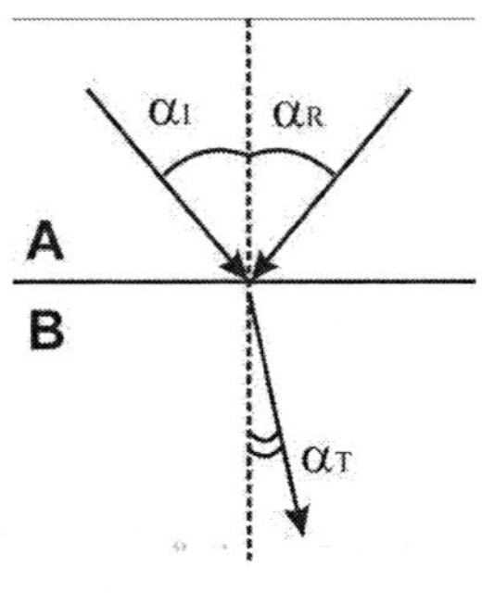

Figure 2. Reflection and transmission at the interface between two media.

In the Eq. (4), α_I and α_T are respectively the angle of incidence and transmission, whereas c_I and c_T are respectively the sound velocity of incident and transmitted waves. The interaction between acoustic waves and media is quite complicated because mode variations may occur under certain conditions. The case of the interface between water and aluminium can be considered as an example. If a longitudinal wave in water impinges on the surface of the aluminium sample, a longitudinal transmitted wave will arise and simultaneously a weak transverse wave appears in the metal. At an angle of incidence of almost 13°, named *first critical angle* for the longitudinal wave, the transmitted longitudinal wave disappears and, in the metal, only the transverse wave propagates. At an incident angle of 20° also the transverse wave in the metal disappears. This last angle is known as the *second critical angle* for the transverse wave. For angle of incidence above the second critical angle, there are no waves propagating in the aluminium sample.

Reflection-transmission phenomenon is responsible for the decrease of the ultrasound wave intensity, but this is not the unique factor affecting the wave propagation within a media. The attenuation of the ultrasound energy can be also due to scattering and absorption which can occur because of the material structure. Generally, the materials are not strictly homogeneous, in fact their structure can include slag inclusions, voids (gas pores), or grains in sintered materials or metallic casts. If the ultrasound wavelength is comparable to the grain size, the interaction between the wave and the material can be assimilated to the geometrical reflection, then the original wave will be splitted into a reflected and a transmitted one. In the opposite case, that is the wavelength is almost 100-1000 times larger than the grain size, the scattering is negligible when compared to the other contribution to the attenuation, but, if the grain size is approximately $1/10^{th}$ of the wavelength, scattering must be taken into account. The *absorption* phenomena can occur due to several causes and it is responsible of the conversion of the sound energy into heat [12]. All these factors can cause the decrease of ultrasounds intensity which follows an exponential law:

$$I = I_0 e^{-\eta x} \tag{5}$$

In Eq. (5), I_0 is the intensity of the wave impinging on the material, I is the intensity at distance x within the material and η is the attenuation factor which depends on the material properties and on every contribution to the attenuation.

1.3. Piezoelectric Effect and Ultrasound Transducers

NDE based on ultrasound methodologies involves the use of different types of probes depending on the inspection technique, the samples shape and size and the type of the investigated defect.

The main part of a ultrasound probe (Figure 3) is the transducer which is made by a piezoelectric material. Therefore US are generated via the Piezoelectric effect, i.e. the property of certain materials to generate an electric potential in response to an applied mechanical stress and viceversa [13]. The first type of transducer was made by quartz crystals, but most of the ultrasonic transducers used today are made by piezoelectric ceramic materials. Indeed, Piezo-ceramics, are characterized by very good transmission properties and can be easily cut in various shapes and size to produce different wave modes.

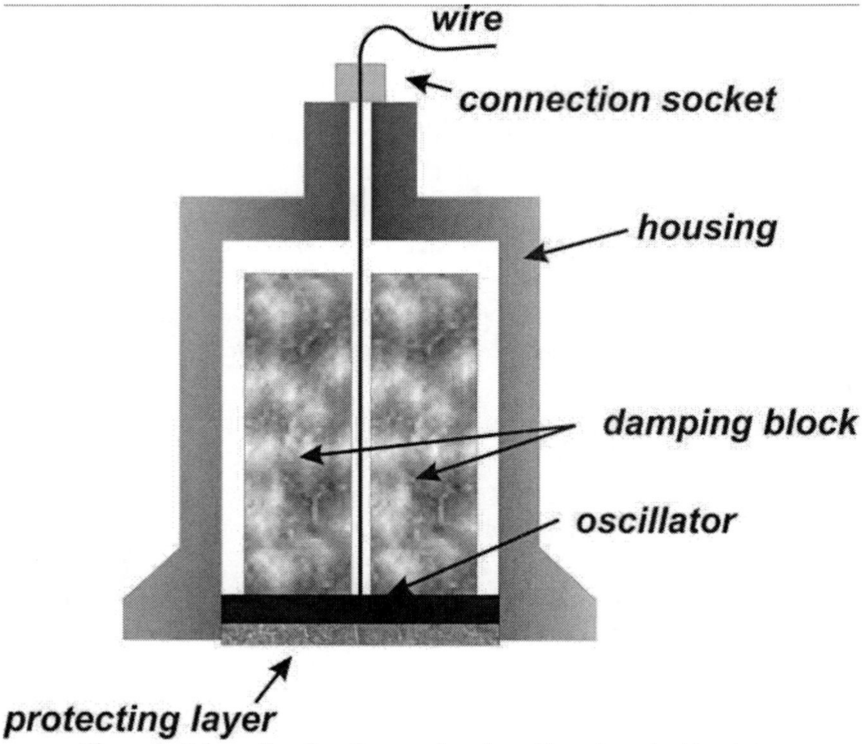

Figure 3. Schematic of a ultrasound probe with a piezoelectric plate.

Such materials also operate at low voltage and are usable up to about 300°C. Barium titanate was the the firstly used piezo-ceramic material, principally for contact transducers, i.e. for tests in which the transducer is in contact with the sample surface. During the 1960's zirconate titanate based compound entered the market [14-15]; such material, at present, is the most commonly employed ceramic for transducers manufacturing. When zirconate titanate transducers are employed in immersion testing, i.e. tests in which both the transducer and the sample are immersed in water, a protection, by water-tight covering layers, is required. Usually the protection is made by a cold-hardening resin which is also used to assure the coupling impedances matching. The transducers surface geometry can be designed in such a way to obtain the ultrasound wave beam focusing, or to obtain a better matching with non flat sample surface. For special applications in which transversal mode wave is needed, the transducers are joined together with a plastic wedge which assures the proper inclination (see Eq. 4) with respect to the plane of incidence of the wave beam.

1.3.1. Transducer Excitation

The transducer excitation is produced by an electrical signal of a specific frequency spectrum, such as tone bursts or pulses (Figure 4). Signals in the form of tone bursts are often used to realize *Trough-Transmission* inspections whereas, for *Pulse-Echo* inspections, a single pulse or a frequency swept is preferred because of the better signal to noise ratio.

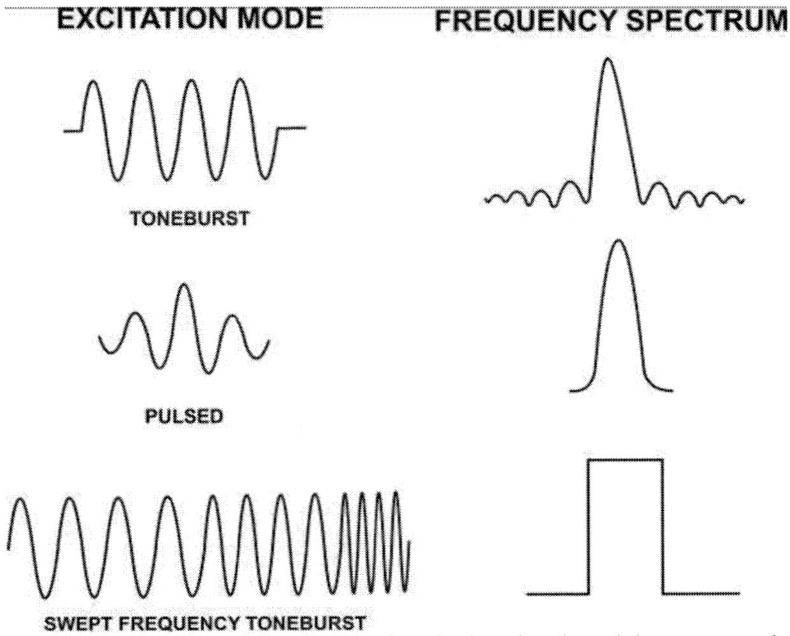

Figure 4. Schematic representation of the most used excitation signals and the correspondent frequency spectrum.

The ultrasound wave beam generated by a transducer has a complex structure: in the *near field*, which is the part most closer to the transducer, the intensity distribution is characterized by maxima and minima because of interference phenomena occurrence. The distance z between the last maximum and the last face of the transducer active element, can be calculated knowing the diameter D of the transducer and the wavelength, then: $z \approx D^2/4\lambda$ (if the wavelength is smaller than the diameter) [11]. *Far field*, the part of the sound field starting from the distance z, generally diverges with respect to the propagation direction. In proximity of the *near field* it is possible to apply the planar wave front approximation.

The higher is the transducer frequency the higher is the axial resolution, but in this case, the penetration distance decreases. Therefore the choice of the frequency is always a compromise between pro et contra. Generally, the transducer capability to detect flaws depends on the combination of several parameters, such as the wavelength, the shape and the diameter of the transducer, the excitation pulse and the nature of the material under analysis in itself. Nevertheless, it is possible to establish some empirical rules to enhance the capability to detect flaws; in fact, it:

- Increases when the flaw size is larger than the wavelength (because scattering phenomena become negligible);
- Increases when the wave beam is focused, i.e. the narrow is the beam the higher is the spatial resolution;
- Increases when the excitation pulse width is narrow (*delta-t* like). In other words, the capability of flaw detection is inversely proportional to the duration of the pulse produced by a ultrasonic transducer. The shorter is the pulse (often at higher frequency), the better is the detection of the defect;

- Decreases in high density materials and/or for high ultrasonic velocity. The signal to noise ratio is inversely proportional to the material density and to the acoustic velocity.

The excitation pulse induced in the transducer and subsequently emitted, can be displayed on an oscilloscope screen (see Figure 6); this pulse is followed by some damped peaks, the higher is the frequency the shorter is the damped part of the signal, which is called *dead zone*.

1.4. Inspection Techniques

To carry out non destructive testing of materials with US, several techniques can be used, but all of them requires the use of a proper coupling media between the transducer and the material under inspection. A proper coupling media allows to maximize the transmitted ultrasonic energy and its efficiency can be estimated through a relation involving acoustic intensity and impedance:

$$\frac{I_T}{I_I} = \frac{T^2 Z_1}{Z_2} \tag{6}$$

In Eq. (6), Z_1 and Z_2 are respectively the acoustic impedance of the coupling media and of the material under inspection and T the transmission coefficient; Eq. (6) gives an estimation of the transmitted intensity I_T if the incident intensity I_I is known. It can be applied if the planar wave approximation is feasible. For example, at an air/carbon-fibre-composite interface, the fraction of transmitted signal is almost 0.03%, whereas if the coupling media is water this fraction increases at almost 70%.

The generated ultrasonic wave is transmitted by the transducer within the sample material through the coupling media, and, as a consequence, the energy is reflected and transmitted at the interfaces.

Two main techniques can be applied to carry out ultrasound testing: *Through-Transmission*, (TT), and *Pulse-Echo*, PE.

To apply the TT, two transducers are used, one acting as *transmitter*, the other acting as *receiver* (Figure 5 a). The transducers are placed respectively on the opposite sides of the sample in such a way the ultrasonic pulses propagated through the sample are received on the reverse side. This technique is based on the measure of the attenuation of the received signal and, the presence of the defects is pointed out by the attenuation variations.

For the PE, only one transducer is required, which works alternatively, as transmitter and receiver (Figure 5 b). Then, informations about the presence of flaws is provided by the reflected signal. To effectively carry out PE, the period of the excitation pulse of the unique transducer must be long enough to avoid overlapping of transmitted and reflected wave.

In both cases, the received signals are converted in electrical signals, and then, which are amplified and displayed on an oscilloscope screen. Then, the informations conteined within the reflected signals can be acquired by a suitable computer software.

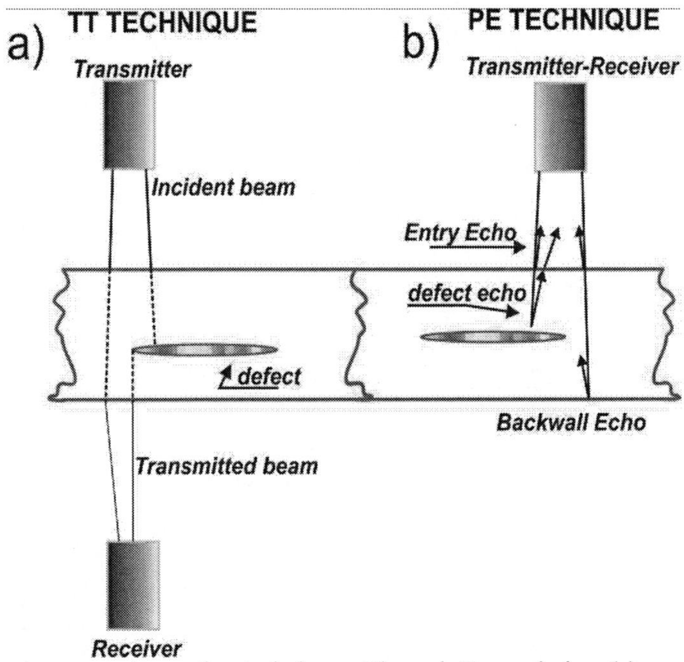

Figure 5.- The most common inspection techniques: Through-Transmission, (a), and Pulse-Echo (b).

1.4.1.Imaging Techniques

The signals received by the transducer during tests, can be analysed in terms of their amplitude and duration. Generally the amplitude of the signals is measured in decibels, whereas the duration is sometimes converted into length. Different imaging techniques are used to highlight a specific response of the material under analysis to the ultrasound waves and to put in evidence the presence of non homogeneous parts or defects. The main imaging techniques, known as A-scan, B-scan and C-scan, are described and discussed in the following paragraphs.

A-Scan

In the Pulse-Echo technique the radiofrequency signal, which is received by the transducer moved point by point on the sample, can be displayed on an oscilloscope screen. In Figure 6 an example of the displayed signal is shown.

To maximize the transmitted signal, the transducer must be oriented with respect to the sample surface, in order to obtain the maximum amplitude of the first echo. It is the so called Entry-Echo (EE), i.e. the peak due to the reflection at the first side of the sample. The last peak displayed is the so called Back-Wall Echo (BWE), that is the signal reflected by the back side of the sample. Every other peak between the EE and the BWE will be due to reflections at intermediate interfaces such as, for example, defects in the material. Thanks to this technique, it is possible to reconstruct the position of a defect. The time necessary to the signal to reach the defect and to turn back to the transducer, which is Δt_2 in Figure 6, is known as *Time of Flight*. If the ultrasound velocity in the material is known, by measuring Δt_2, the distance of the defect from the surface can be calculated. When the ultrasound beam runs across a defect it is partly reflected and partly transmitted and the signal undergoes an attenuation; in general, the larger is the reflected signal the smaller will be the BWE.

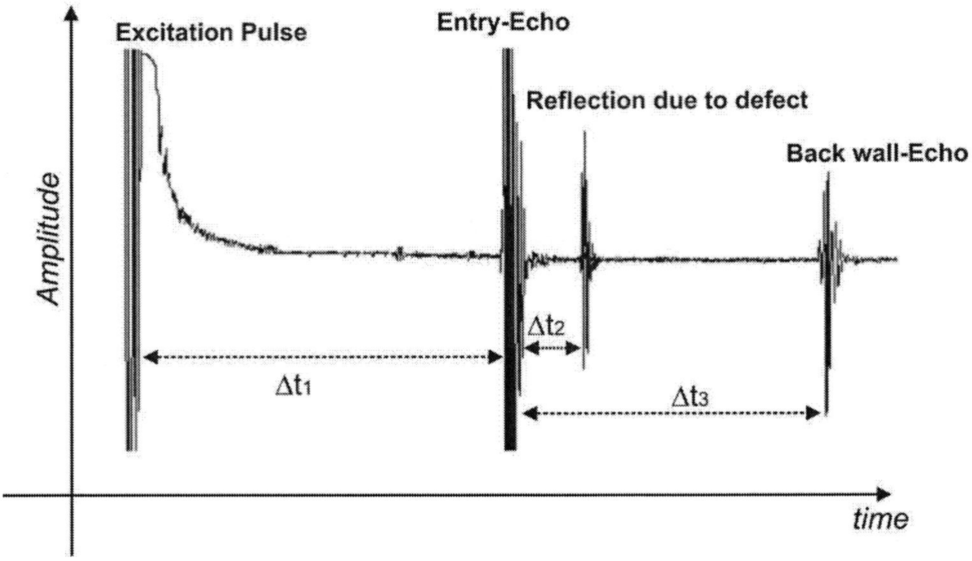

Figure 6. Typical A-scan representation.

It is important to notice that the defect orientation plays an important role in this kind of testing, in fact, although the defect is large, it may be not detected if it is oriented parallel with respect to the ultrasound beam propagation. Moreover, if the defect has a smaller impedance (for example it is an air bubble) and it is larger than the wavelength, a mirror-like reflection can occur and, as a consequence, the BWE can disappear.

The A-scan can be used to estimate the acoustical attenuation, in a material, because it is related to the amplitude decreasing of repeated echoes of the signal [16]. In the A-scan imaging technique, the axial resolution depends on the wavelength and on the pulse duration. The spatial resolution depends on the distance between the transducer and the sample surface, therefore, it depends on the divergence of the beam and on the ratio between the transducer diameter and the wavelength.

B-Scan

By moving the transducer with a traversing system, is possible to acquire A-scans point by point along a line (Figure 7). All those information can be displayed in a graph in which the geometrical transducer position is on the *x* axis and the time delay on the *y* axis. This representation is known as B-scan. To each point on the graph is associated a signal intensity attenuated by the reflections or attenuation phenomena occurred in that point. Thanks to the B-scan not only the position of a defect can be detected but also, its extension along the scanning line is provided. The axial resolution is the same as for the A-scan, whereas the spatial one depends also on the step of the scanning (i.e. the distance between two successive points).

C-Scan

Another very useful imaging technique is the C-scan. In this case the transducer is guided to move along a pattern covering all the sample surface, as sketched in Figure 7 b.

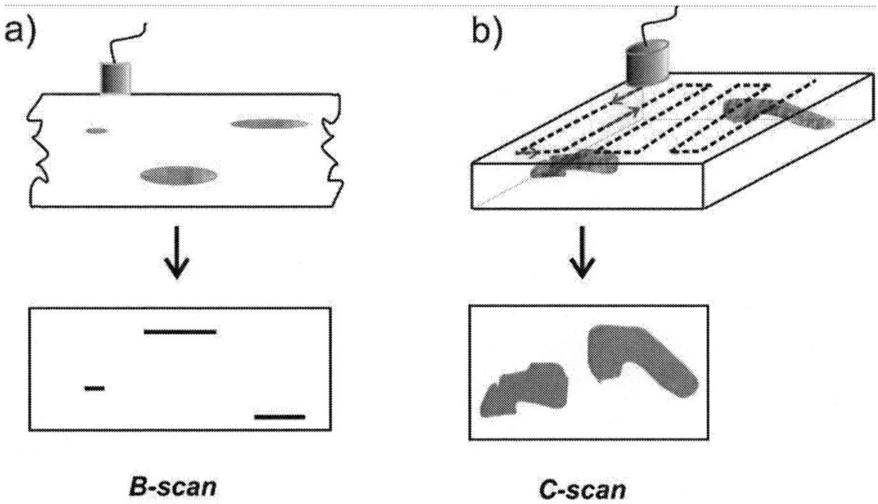

Figure 7. A schematic of B-scan, (a), and C-scan, (b), imaging techniques.

To carry out C-scans, the transducer is often moved by a traversing system which "communicates" its position to a computer, as well as, the signal received by the probe.

Some detectors have *gates* whose height and duration is fixed or, in certain systems, it can be chosen by the user. The gates are similar to "time windows" in which the acquiring system works, therefore just only the most interesting part of the signal can be monitored. When the PE technique is used, by adequate gates positioning, only some parts of the thickness can be inspected and the dead zone of the signal can be avoided.

A gate placed on the BWE may be used to monitor the attenuation undergone by the signal after passing through the whole sample, and, in some systems, the amplitude of the gate can be used as threshold to detect just only the defects which can not be accepted in a "damage tolerant" approach (see paragraph *2.3.1*).

The information acquired in a gate can be represented as C-scan in which, the in-plane extension of a defect is supplied as the projection of the defects on a plane (see Figure 7 b). A C-scan can be an amplitude map of the received signal or a *Time of Flight* map. This type of imaging technique has the same characteristics as the B-scan concerning both the axial and the spatial resolution.

2. Ultrasounds in the Aeronautical Field

Since the first ultrasound application, the primary scope was to have a tool to detect defects in materials and structures due to the manufacturing processes. Therefore, the first role of ultrasonic inspection, as well as for other non destructive inspection methodologies, was to assess the quality of the products in order to comply with the design requirements. The need to assess the re-usability of service-aged structures, made, in a few time, ultrasounds became also an important tool to search for damages produced during the in-service life of components.

In the aeronautical industry, ultrasounds are considered the most efficient and widely used inspection method at the production line to precisely identify the flaws whereas other techniques such as shearography, Infrared thermography, and tap-check are considered appropriate for an overall control of large components [17].

Typically, ultrasonic inspection is carried out to assess the quality of metallic components in order to find internal defects due to the manufacturing process such as voids, inclusions, discontinuities, cracks or defect in the assembled structures which were obtained by welding [18-19] or bonding joints, complex parts of aircraft engines, in riveted parts of the fuselage etc.

During the 1980's carbon fibre reinforced plastics (CFRP) have been introduced to replace specific metallic components in order to reduce weight, to facilitate realization of complex shape parts and to decrease production costs. In the 2000's Airbus launched the production of the A380 which is the first aircraft with a wide use of composite materials (25% by weight) such as CFRP, glass-fibre reinforced plastic and aluminium-glass-fibre laminate. In the A380, composite materials are extensively used in wings, fuselage sections, tail surfaces, doors and the central wing box is entirely made of CFRP.

Since the first applications, composites presented different problems, when compared to metallic materials, in terms of type and occurrence of flaws and in-service life behaviour. This led to the need of new examination processes, such as more accurate ultrasonic inspections, to provide reliable responses to the severe certification requirements.

Being composites obtained through complex manufacturing processes involving different materials, a significant dependency between mechanical properties and processing procedures exists. Therefore the introduction of defects during the manufacturing must be carefully investigated.

Moreover carbon fibre composites are particularly sensitive to accidental impact damages during their in-service life due to the combination of low ductility and low mechanical properties in direction orthogonal to the lamination plane of the fibre.

Together with flaw detection, the repairing techniques are becoming an important topic to extend structures life, thus leading the ultrasound inspection to become fundamental for the assessment of the quality of the repaired components and the process procedures.

In the following paragraphs composite manufacturing techniques will be briefly recalled in relation to the typical induced defects, whereas impact damage will be discussed as one of the principal cause of failure of composite structures. Moreover most widely used setups and ultrasonic testing techniques in aeronautics will be presented.

2.1. Composite Materials Manufacturing

The words composite material concern on structures made by different materials.

Fibre composite materials, that are the most frequently composites adopted for aeronautics structures, are manufactured from a relatively homogeneous and weak material (*matrix*), like epoxy resin, and strong fibrous materials like carbon or glass fibres (*reinforce*). Fibres are mainly demanded to carry the load whereas the matrix keeps the fibres together, distributes the load and protects the fibres from the environment.

Fibre diameters can range from 1 μm up to 10 μm and, generally, their length considerably exceeds the diameter.

Composite materials for structural applications are often manufactured in thin layers. Structural components are made by stacking sequences of layers (laminate) to obtain the prescribed thickness and the required mechanical properties. The desired mechanical properties are obtained also by selecting the fibre orientation of each layer and the staking sequence of the laminate.

Some of the most common processes to manufacture fibre composite materials are:

- *Wet lay-up*

Wet lay up is a manual manufacturing process: on a mould prepared with a gel coating a series of properly oriented carbon fibre lamina are overlapped and, on the top, viscous resin is distributed by a brush. The resin in excess is squeezed out by a roller and it is cured at room temperature, although by suitable heating systems, the process may be accelerated. The composite components obtained presents one side with smooth surface and one with rough surface. This technique is cost effective and requires trained personnel because the quality of the products depends not only on the employed materials but also on the ability of the operator. The wet lay-up is not suitable for the production of large parts.

- *Hand lay-up*

This is the most used technique in the aeronautic industry. It is similar to the previously described process, but it involves pre-preg materials, i.e. fibre lamina which already contains an amount of the matrix material used to bond them together. To obtain the final product, a vacuum bag is applied and autoclave curing is performed. Pre-pregs are costly and need to be stored in a cold chest at –18 °C all before use. For these reasons, hand lay up is mainly employed for the production of very special parts for aerospace and automotive industries and also for the production of structural parts thanks to the good fibre-volume percentage (higher than 50%). However, until 2000, most component of Boeing airplanes have been made by hand lay up.

- *Resin Transfer Moulding*

To carry out Resin Transfer Moulding, RTM, two matched metallic moulds are required. A series of dry fibre plies are settled in the mold and kept together by stitching or by the use of binders. During the process the moulds are clamped together and heated as well as the resin in order to reduce its viscosity. After that, the resin is injected in the mould in which vacuum is performed. The main advantages of RTM are the composite dimensions only depend on the mould shape and pressure vent or other devices can be slot in by using more complex moulds. This process is cost effective when a large amount of components have to be produced and the manufacturing of thicker parts need less time than in other production processes. Moreover, the RTM products are ready to use whereas parts made by hand lay-up need further treatments. The obtained composites have good surface finish on both sides which is an advantage for ultrasound inspection.

This technique is used in the production of a range of many components including door pillars, stringers, stiffened panels, rudder tips and ribs.

- *Resin Film Infusion*

This process involves just one male or female mould of the desired shape. A thin film or sheet of pre-catalyzed solid resin is laid into the mould and the pre-form (in the shape of the desired part) is laid over the resin film. A vacuum bag and sealant tapes close the mold. The resin is thermally cured by an oven or autoclave while a vacuum pump is applied. This process allows obtaining materials of high specific strength and it's easier to carry out than hand-lay up but it is not used for manufacturing of structural parts. Radar domes and the Airbus A380 [20] rear pressure bulkhead are made by RFI.

- *Resin Infusion*

This process is well known since the sixties, when it was hardly employed for yacht part production; nowadays it is used in the manufacture of large-scale composite structures for aerospace and transportation applications.

It is similar to the Resin Film Infusion, but in this case, at the beginning there are only dry fibres in the mould.

All the air in the "vacuum bag" is evacuated with a vacuum pump causing compression of the dry stack of reinforcement fabrics. Through a series of feed hoses sealed into the bag, liquid resin is then sucked via the vacuum from large mixing containers and then it migrates throughout the mold saturating the entire stack of laminate. The vacuum is kept on for several hours until the resin has cured.

This way of fabricating composite parts is becoming popular for its own merits. Resin infusion is far more cost effective than the pre-preg handing-up. This is particularly suitable for fabrication of large and thin parts. A limitation of this method requires resins of low viscosity which may lead to the composite part which may be brittle and of poor impact resistance but this can be overcome by proper design.

2.2.1. Main Types of Defects in Composites Induced by Manufacturing Processes

Defects in composites may arise during production, because of the several different parameters, such as pressure, temperature etc., that are involved in the production processes. Dimensions and density of the defects are typical of the specific process [21].

Some of the most common defects are:

- Voids
- Slag inclusions
- Incorrect fibre-volume fraction
- Bonding defects
- Fibre misalignment

The detection of voids is particularly crucial because this type of defect can reduce the inter laminar shear strength which is responsible of delamination occurrence (inter lamina debonding). Therefore severe degradation on strength and stiffness of the component in-service [22] can be expected. In manufacturing processes like hand lay-up, RTM and RFI the voids formation depends on different parameters, such as the fibre volume fraction (the fraction of the volume occupied by fibres), fibre permeability and resin viscosity. It has been shown that void formation in RTM is directly related to imperfections on fibre arrangement which leads to a non-uniform permeability [23]. As a consequence of this inhomogeneity, the resin flow is perturbed and then void presence can occur. Stitched stiffened composite plates obtained by the RFI process can present voids because of an non optimal two-dimensional resin flow inside the mould. Moreover, the stitches location can also induce void formation, as they somehow block the resin flow [24].

The incorrect fibre volume percentage, due to an inadequate resin content, causes a weakening of the structure [25]. Often, it is also associated with the presence of voids which are related to an incorrect resin curing process.

Delaminations can be usually related to thermal effects and especially when the produced part is thick (over 10 mm). Inclusions can be things such as excess binder if it is hand-applied, metallic parts from cutting tools, release agent and waste resin from the previous moulding.

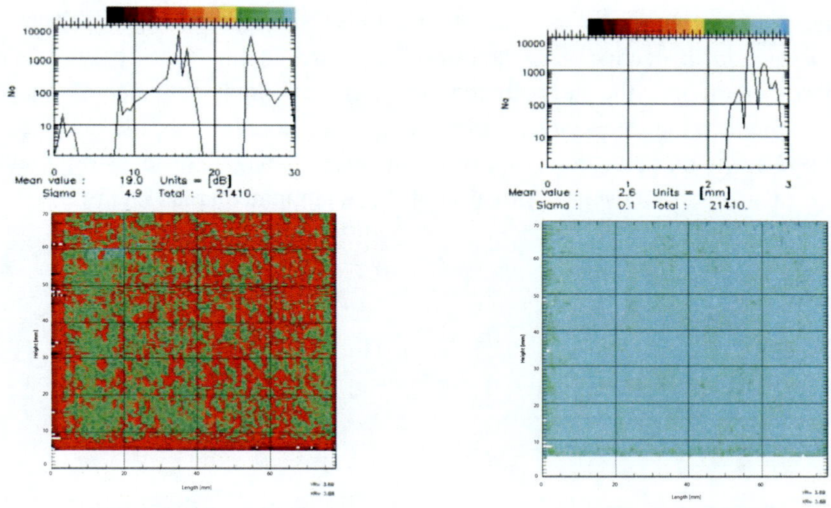

a) *Amplitude C-scan* **b) *ToF C-scan***

Figure 8. Amplitude (a) and Time of Flight (b) C-scans obtained applying Pulse-Echo technique on an hand lay-up Carbon Fibre Reinforced Plastic Composite. In the histogram above each C-scan, respectively, the number of points vs. amplitude and the number of points vs. time of flight (as lengths measurement) are reported. In the (a) part of the figure the green area highlight the presence of porosity whereas in (b) constant thickness is noticeable.

Most defects can be quickly detected by the TT technique whereas the PE can provide more detailed results and it is generally preferred to detect microporosity around stitching, displaced fibres and fibre volume fraction changes.

PE is very effective in fibre volume fraction estimation in RTM composites: every variations of the time of flight are due to fibre-volume fraction variation since the manufacturing process assures a constant thickness of the produced component. Variations of the fibre or resin content determine a local impedance variation with respect to the coupling media impedance. As a consequence, the ultrasound velocity is modified producing a variation of the time of flight.

The attenuation measurement could be a sufficient and effective measurement to detect areas affected by gross porosity or voids [26]. In Figure 8, an example of the effect of porosity is noticeable: the porosity is recognizable through to the inhomogeneity of the amplitude in the a-part of the figure, whereas, in the b-part of the figure, the ToFs converted in lengths (mm), give no evidence of the defect, because no thickness variations are read.

2.2.2. Defects Induced by Impacting Objects

In the design of aeronautical structures, impacts are a problem commonly known and treated. Accidental impacts can occour during aircraft takeoffs and landings, when stones from the runway are lauched at high velocities against the structures, as well as during manufacturing and maintenance when tools could drop on the components. In the last case impact velocities are small but the mass of the impacting object may be large.

In laminated composite structures, impacts inducing a barely visible indentation on the surface can often create significant internal damages. In that cases, a severe reduction in strength can be induced and the damage can grow under load. UT, is the most utilized non destructive evaluation methodology to detect the impact damages. In the case of low energy impacts, the damage can consists of delaminations, matrix cracking and fibres failures. Typically when impact occurs on a laminated structure, at interfaces between layers with different fibre orientation, the delaminated area has an oblong or "peanut" shape with its major axis oriented in the directions of the fibre in the lower layer at the interface [27]. The UT provides C-scans that are projection of all of these damaged surfaces on a single plane, which can be measured to have an estimation of the delaminated area (Figure 9).

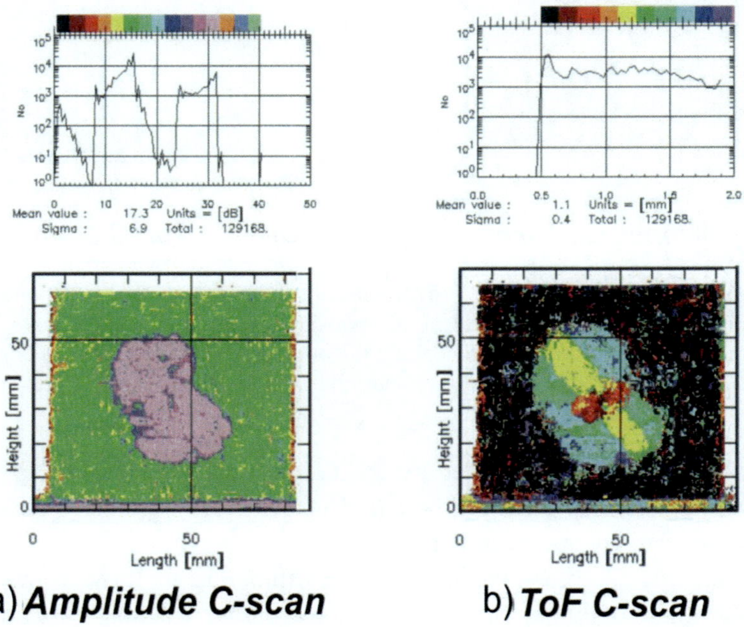

Figure 9. Amplitude (a) and Time of Flight (b) C-scans obtained applying Pulse-Echo technique on an hand lay-up Carbon Fibre Composite. In the histogram above each C-scan, are reported respectively the number of points vs. amplitude and the number of points vs. time of flight. In (b) the ToF has been converted in lengths so the typical oblong delamination are noticeable as length variations: each colour corresponds to a different distance from the surface.

2.3. Inspection of Aeronautical Structures

Industrial inspection procedures prescribe the use of reference standards [28] for calibration of ultrasonic equipments in order to carry out effective measurements. Reference standards allow to obtain flaw measurements as reliable as possible and help to interpret and quantify the information obtained. Reference standards assure repeatability of measurements also when different equipment are used. There are different reference standards for ultrasonic inspection in metals, whereas for inspection of composites suitable references with simulated defects must be specifically manufactured.

Generally, the core of a ultrasonic testing setup is the transmitter-receiver unit able to excite the transducer, to receive the electronic signal and to amplify signals which can be displayed on an oscilloscope screen. In the last decades different setups have been realized for the aeronautical industry because the different application requirements and the faster rate of inspection needed.

2.3.1. Portable Systems

The need to inspect large components as quickly as possible is one of the main concern in the aeronautical industry. In most cases portable systems are adopted for in-situ inspection during in-service life of the components [18, 29-30].

The oldest kind of in-situ inspection system involves just one transducer which is manually driven to get a point by point A-scan. In some equipments an amplitude gate can be used to point out reflections with amplitude over/below a certain threshold. When this occurs, a flash light or a sound is emitted. One of the most common utilized criteria is to search for points where the amplitude of the signal is halved (a 6 dB drop), then all those points are marked with a pencil and, in certain cases, they undergo a more accurate inspection such as immersion testing. These kind of flaw-detector are widely used to find discontinuities in welds, forgings, billets, axles, shafts, tanks and pressure vessels, turbines, and structural components but they do not allow for measuring of the defects size and, as a consequence, of their extension.

More advanced portable systems are equipped with scanning arms which monitor the transducer position and its output. They provide A-scan and C-scan on time and allow to zoom on the suspected areas, therefore defect dimensions can be reconstructed. This kind of equipment can also have fully automated scanning systems with computer data storage. To inspect very large component of complex geometry (such as fuselage parts) the transducer can be moved by crawling robots allowing the possibility to interact within the automated scanning process.

2.3.2. Ultrasound Immersion Testing

Ultrasound immersion testing is one of the most used technique to inspect composite materials. For their faster rate of inspection and the possibility to analyse very large components they are often the final step of the production process. Ultrasound immersion testing allows to obtain A-scan, B-scan and C-scan, it is simple to use and can be easily automated.

To apply this technique both sample and transducers are immersed in water which is the coupling media. The transducers during the scanning are moved by mechanical arms so their orientation and their distance from the sample are known. When the PE technique is used, the transducer is headed perpendicularly to one side of the sample. If the Double Through-Transmission technique is used, i.e. a special kind of the PE testing, a glass mirror is put closed to the opposite side of the sample. Because of the very large impedance difference, the glass mirror may be considered as a perfect reflector. In this way, the signal passes through water, sample material, water and *viceversa*. Knowing the distance of the mirror from the sample surface and the ultrasound velocity in water the contribution to the attenuation related to the sample can be deduced. By using the Double Through-Transmission technique the sample thickness and the ultrasound wave velocity in the material can be measured [31-32].

Furthermore the mirror allows obtaining separated echoes and more precise information about the interfaces position. In immersion inspection the transducer must be precisely positioned orthogonally to the sample surface that is usually achieved by the computer controlled mechanical arm.

In general, the transducers for immersion testing are planar, i.e. the active element is cut with planar and parallel sides, producing divergent wave beams. In this case it is preferable to work within the *near field* limit. There are also focused transducers in which the active element has a concave face, in order to maximize the signal intensity in the focus spot. Those transducers have higher sensitivity and spatial resolution than the planar ones. Focused transducers are therefore suitable for the inspection of very small components or when an accurate measurement of the defect dimensions is required.

To carry out ultrasound immersion testing a water pool is required and, depending on the sample dimensions, the pool may be often quite large. Another very useful way to carry out ultrasonic testing with water coupling is based on using jet probes (*squirters*). They are transducers provided with a laminar water jet which impinges on the sample surface acting as a wave guide for the ultrasound beam, so, the water pool is not needed. Water jet probes are widely used to carry out TT testing allowing very quick inspection of large components.

3. New Concepts for Ultrasonic Inspections

During the last ten years new ultrasound based instruments and inspection methodologies have been developed. In the following two of the most successful solutions are presented.

3.1. Ultrasonic Arrays

The simplest form of a multi-element ultrasonic array for non destructive testing is a series of several single element transducers arranged in such a way as to increase inspection coverage and/or speed of inspection.

The functioning of an array requires a system for multiplexing electronics to excite each element of the array in turn and continually cycle through the elements. Each element has a separate pulser-receiver with individual gain and damping controls for adjustment of ultrasonic intensity. The elements are much smaller than conventional transducers, and can be pulsed as a group or in a way to generate directionally controllable wavefronts. Wavefront shaping is obtained by conveniently exciting every element of the array, in order to introduce controlled delay among the elements. These kind of arrays are known as Phased Arrays. Phased arrays allow to both steer and focus the sound beam through a range of angles and focal depths, resulting in large improvements in productivity and reliability for inspections.

While phased array transducers come in a wide range of sizes, shapes, frequencies, and number of elements, what they all have in common is a piezoelectric element that is divided into a number of segments. Contemporary phased array transducers for industrial NDE applications are typically made by piezo-composite materials, which are made up of many tiny, thin rods of piezoelectric ceramic embedded in a polymer matrix.

Although Phased Arrays for CFRP inspection were introduced on the market some decades ago, since they provided only B-Scan representation, which was unfamiliar to most of the users, they started being widely used only in the last years. In particular the possibility to manage the beam focusing is becoming very attractive for industrial inspections [33]. Nowadays arrays are available with up to 128 elements with a spacing among successive elements as small as 0.25 mm [34]. Sometimes the phased array probe is assembled with a plastic wedge in order to assure coupling in both shear wave and longitudinal wave applications.

3.2. Non-Contact Ultrasonic Inspections

One of the principal requirement for effective ultrasound inspection is the use of a proper coupling media. Depending on the applied technique and on the chosen transducer, water, gel or oils can be used as coupling media. In some cases these substances can contaminate the tested parts (especially aerospace components) then the use of air as coupling media is nowadays one of the future challenge in ultrasonic inspections.

Air Coupled Ultrasounds

Over the last 15 years new ultrasound probes for air coupled testing have been developed but some crucial problems remain. In fact, bearing in mind Eq. (5), using air as coupling media a lot of ultrasonic energy is lost during transmission and the higher is the frequency the larger is the attenuation by the air.

To overcame such problems the air coupling is generally used in TT configurations. At present, the transducers are designed for low-frequency applications (30 Hz up to 100 Hz) or for medium frequency applications (0.5 MHz up to 3 MHz). Low frequency transducers provide a better signal to noise ratio, but recently it has been proven that medium frequency transducers are able to perform composite materials inspections [35-36].

Laser US

Laser ultrasound is based on the physical principle of thermal expansion (also called thermoelastic regime) or ablation. In the thermoelastic regime, the ultrasound is generated by the sudden thermal expansion due to the heat shock of a tiny surface of the material by the laser pulse. If the laser power is enough to heat the surface above the material boiling point, some matter is evaporated (typically some nanometers) and the ultrasound is generated by the recoil effect of the expanding material. In the ablation regime, plasma is often formed above the material surface and its expansion can make a substantial contribution to the ultrasonic generation.

Laser ultrasounds can be used as generators as well as detectors in the materials inspections. The effectiveness of this technique has been proven for metals but nowadays it is used also when a non-contact inspection of composites is required [37-38]. The produced ultrasound can be detected by a transducer or by interferometrics techniques, taking advantage of the superficial deformation induced in the sample by the incoming ultrasound wave. Most industrial laser-ultrasonics applications use confocal Fabry-Perot [39-40] or photorefractive interferometers to demodulate the detected light.

Most common lasers used for ultrasound generation are solid state Q-Switched Nd:YAG and gas lasers (CO_2 or Excimers) because of their short pulse (from tens of nanoseconds to femtoseconds) and high peak power. Detection lasers are continuous or long pulse (typically of tens of microseconds) and with long coherence length.

3.3. Advantages Versus Disadvantages of Ultrasonic Inspections

As every non destructive testing method, also Ultrasonic testing has its *pro and contra*. Some of the main advantages of ultrasonic inspection can be resumed as follows:

- Both surface and bulk discontinuities can be detected and the high sensitivity allows the detection of extremely small flaws;
- In general the depth of penetration for flaw detection is larger than the penetration obtained by applying other NDE methodologies;
- Only one side access is required in Pulse-Echo inspections;
- Capability of estimating the size, orientation, shape and nature of defects and, the position and the size of the defects are measured more accurately than using other NDE methods;
- Minimal part preparation is required;
- Ultrasonic setups provides instantaneous results;
- Possibility of portable or highly automated setups producing detailed images;
- It has other uses, such as thickness measurement, in addition to flaw detection;

On the contrary, main disadvantages includes:

- At least one surface of the part need to be accessible to transmit ultrasound
- Extensive technical training and skilled personnel is needed
- Traditional Ultrasonic testing requires a coupling medium to assure the sound energy transmission through the part therefore, inspected items must be water resistant, when using water based couplants;
- High dependence on the surface and the geometry of the components: materials that are rough, irregular in shape, very small, exceptionally thin or not homogeneous are difficult to inspect;
- Cast iron and other coarse grained materials are difficult to inspect due to low sound transmission and high signal noise ratio;
- Linear defects oriented parallel to the sound beam may go undetected;

- Reference standards are required for both equipment calibration and the characterization of flaws.

CONCLUSION

The introduction of composite materials in the aeronautical industry has emphasized the role of the NDE methodologies within the production processes. NDE allows not only to asses the accomplishments of the quality design requirements but also to monitor components during in-service life. In this scenario, the ultrasound based method has taken a leading role thanks to its effectiveness and its ability to provide accurate and reliable measurements of defect position and size. Once the type of the investigating defect has been individuated, by choosing the proper technique and set up, fast inspection of large components and/or more accurate defect characterization can be achieved. Defects such as porosity, delaminations, slag inclusions in composite structures may be effectively detected, using Through-Transmission and/or Pulse-Echo technique, and displayed through imaging techniques such as A, B and C-scan. In particular B-scan is becoming very useful thanks to the introduction of phased array transducers which allows to manage the wave front and the focus of the probing ultrasound beam. New transducers design or the use of different ultrasound sources has allowed the inspection of components using air as coupling media, i.e. to carry out inspections without any invasive interaction with the inspected components.

From the technological point of view, since 1950's until today, significant improvements were achieved resulting in smaller and more advanced instruments of great capabilities. The introduction of LED or LCD screens in place of the previous cathode ray tubes, have made possible the vision in different and sometimes difficult light conditions. Modern computers together with motion control and robotics have contributed to the advancement of ultrasonic inspections: large part and/or complex shape components inspections and multiple transducer information can be programmed, managed and processed by suitable computer software.

Nowadays, during the design process, Quantitative Non Destructive Evaluations has become a tool as important as failure-related engineering disciplines, nevertheless, it is impossible to avoid manufacturing flaws and in-service material damaging thus a continual improving in flaw detection and in characterization techniques performances is necessary.

REFERENCES

[1] Richardson, L. British Patent No. 9423, 1912
[2] Langévin, P.; Chilowsky C., French Patent No. 502913, 1916
[3] Sokolov, S.Y., *Zavodskaya Laoratoriya* 1935, *4*, 1468-1473
[4] Pohlman, R. Z. *Phys*. 1939, *113*, 697-709
[5] Firestone, F. *J. Acoust. Soc. Am.* 1946, *17*, 287-299
[6] http://www.library.uiuc.edu/archives/archon/index.php?p=collections/controlcardandid=4662
[7] Dussik, K.T. Z. *Neurol. Psychiat*. 1942, *174*, 153-168

[8] Ludwig, G.D.; Struthers, F.W. Naval Medical Research Institute Reports, Project #004 001, Report No. 4, June 1949
[9] McNay, M.B.; Fleming, J.E.E., *Ultrasound in Med. and Biol.*, 1999, *25*, 3 – 56.
[10] Achenbach, J.D. in Proc. *AIP* Conf., 2009, *1096*, pp.3-15.
[11] Ktrautkrämer, J.; Ktrautkrämer, H., *Ultrasonic Testing of Materials*; Springer-Verlag: Heidelberg, DE, 1990, 4thedition
[12] Mason, W.P. *Physical acoustics and the properties of solids*; Van Nostrand: New York, 1958
[13] Lippman, G. *Annales de chimie et de physique* 1881, *24*, 145
[14] Jaffe, B.; Roth, R.S.; Marzullo, S. *J. Appl. Phys*. 1954, *25*, 809-810.
[15] Mattiat, O.E. *Ultrasonic Transducer Materials*, Plenum Press, New York, US, 1971
[16] Roderick, R.L.; Truell, R. *J. Appl. Phys.* 1952, *23*, 267-279.
[17] ASTM E2533 - 09 Standard Guide for Nondestructive Testing of Polymer Matrix Composites Used in Aerospace Applications
[18] Berke, M.; Kleinert, W.D.; Reinersmann, J.; Schlengerman, U.; Volkmann, K. in Proc. of the 15th World Conference on *Non-Destructive Testing*, October 15-21, 2000, Rome, Italy
[19] Al-Nuaimy, W.; Zahran O., in Proc. of the European Conference on *Non Destructive Testing,* Sept. 25-29, 2006, Berlin, Germany
[20] Black, S. *An Elegant Solution For A Big Composite Part*, 2003
[21] http://www.compositesworld.com/articles/an-elegant-solution-for-a-big-composite-part.aspx
[22] Adams, R.D.; Cawley, P. *NDT Int*. 1988, *21*, 208-222.
[23] Yoshida, H.T.; Ogasa, T.; Hayashi, R. *Comp. Sc. and Tech*. 1986, *25*, 3-18.
[24] Kang, M.K.; Lee, W.I.; Hahn, H.T. *Comp. Sc. and Tech* 2000, *60*, 2427-2434.
[25] Han, N.L.; Suh, S.S.; Yang, J-M.; Hahn, H.T. *Composites Part A* 2003, *34*, 227-236.
[26] Armanios, E.A. *Composite Materials*: *Fatigue and Fracture*, STP1285 Sixth Volume; ASTM: West Conshohocken, US, 1997
[27] Stone, D.E.W. *RAE* Tech. Report, 74162
[28] Serge, Abrate, *Impact on composite structures*, Cambridge University Press: Cambridge, UK., 1998
[29] ASTM E428-08 Standard Practice for Fabrication and Control of Metal, Other than Aluminum, Reference Blocks Used in Ultrasonic Testing
[30] Dragan, K.; Klimaszewski, S. in Proc. of European Conference on *Non Destructive Testing*, 25-26 Sept., 2006, Berlin, Germany
[31] Sánchez, J.M.; Ojeda, F.; Ávila, C.; Borrás, M.; Aldea, J. in Proc. of European Conference on *Non Destructive Testing*, 17-21 June, 2002, Barcelona, Spain
[32] Bashyam, M. in Proc. Of the 17th Annual *Review of Progress in Quantitative Nondestructive Evaluation* La Jolla, Plenum Press, CA, 1991, 10A.,1035-1042
[33] Roth, D.J. NASA TM-107184
[34] Willsher, S.J.; Smith, R.A. in Proc. of *NDT '98*, The British Institute of NDT, 1998, 46-49
[35] Rau, E.; Grauvogl, E.; Manzke, H.; Cyr, P. in Proc. of the European Conference on *Non Destructive Testing*, Sept. 25-29, 2006, Berlin, Germany
[36] Schindel, D.W.; Hutchins D. A. *Ultrasonics* 1995, *33*, 11-17.

[37] Gan, T.H.; Hutchins, D.A.; Billson, D.R. in Proc. *IEEE Ultrasonic* Symposium, Munich, Germany, 2002, pp. 875-878.
[38] Monchalin, J.P.; Neron, C.V. *Canadian Aeronautics and Space Journal*. 1997, *43*, 1
[39] Stratoudaki, T.; Edwards, C.; Dixon, S.; Palmer, S.B. in Proc. of European Conference on *Non Destructive Testing*, 17-21 June, 2002, Barcelona, Spain.
[40] Walter. J.B.; Telschow, K.L. in Proc. *AIP* Conf 2000, *509*, pp. 247-254.

In: Recent Advances in Non-Destructive Inspection
Editor: Carosena Meola, pp. 59-88
ISBN 978-1-61668-550-8

Chapter 3

ULTRASONIC AND SONIC TECHNIQUES APPLIED TO CONCRETE AND MASONRY STRUCTURES

Vincenza A.M. Luprano* * *and Camilla Colla °
* ENEA S.S. 7 Appia km 714 - 72100 Brindisi, Italy
° DISTART Department, Engineering Faculty, Bologna University, v.le Risorgimento 2, 40136 Bologna, Italy

ABSTRACT

The quality assurance during and after construction of new structures and after refurbishments, the characterisation of material properties and damage due to time and environmental influences are more and more becoming a serious concern. In recent years, innovative Non Destructive Testing (NDT) techniques, applicable for the assessment of existing civil structures, have become available for in situ analysis on concrete and masonry structures, but they are still not established for regular inspections. In addition, in the civil engineering field, repairing and anti-corrosion protection of the building structure is one of the most important applications of polymer composites. Nonetheless, the adhesion between overlays and concrete or masonry substrate is a concerning factor affecting the reliability and durability of repair. During the change of environmental conditions (temperature and humidity) after curing, the need exists for adhesion monitoring in the case of large area objects like industrial floors, bridge decks, injected concrete structures. In this chapter two main objectives will be discerned. The first relates to the application of sonic and ultrasonic testing techniques to measure in situ the development in time of the mechanical properties of concrete and polymer composite. The second one deals with the assessment of damage and defects in concrete or masonry, and between concrete or masonry and applied polymer composite by means of sonic and ultrasonic techniques. Different data acquisition procedures and methodologies together with example applications will be described including tomography, scanning acoustic microscope, transmission and surface measurements. Further, an impact–echo example application on post-tensioned concrete will be described. The importance of data visualisation in form of 2-D images will be discussed. All these are monitoring techniques that can be applied in situ after their validation in laboratory.

1. Introduction

In recent years, innovative Non Destructive Testing (NDT) techniques and procedures, applicable for the assessment of concrete and masonry materials, have become available for on-site use on existing civil structures, bridges and dams [1-7]. Nonetheless, their use is still not established for regular inspections. Efforts in the normative field have been attempted in the last years to write guidelines for the assessment of in situ compressive strength in structures and precast concrete components [8] using traditional non destructive and semi destructive methods such as ultrasonic wave, rebound number, penetration resistance, pullout test and others.

The use of ultrasonic techniques for the assessment of concrete properties as recommended by standards in UK, USA, EU, Germany, and by RILEM, is well reviewed in the paper by Komlos [10]. Furthermore, studies have been performed on ultrasonic assessment of early-age changes in the properties of cementitious materials by RILEM's TCs.

Although ultrasonic NDT is widely employed and investigated, its applications to structures is still highly demanding on user's skill. The aim of this chapter is to illustrate traditional and innovative non-destructive sonic and ultrasonic techniques in the assessment of existing civil and historic structures, also through case studies.

2. Principles of Acoustic Techniques

2.1. Ultrasonic Technique

The propagation of elastic waves at high frequency (higher than 20 kHz), as in the ultrasonic range, corresponds to a dynamic deformation in the material. This deformation produces a displacement around the equilibrium position of atoms and chain segments in the nm range, so the information that can be obtained from the ultrasonic measurements is related to small-scale mobility of short segments between entanglements. On the other hand, the deformation rate is very high since the typical employed frequencies range from the tens of KHz to the tens of MHz. When an alternating time-dependent deformation at ultrasonic frequencies is applied to a material, the propagation velocity of elastic waves (ultrasonic or acoustic velocity) and the attenuation of the waves amplitude (called ultrasonic or acoustic attenuation) can be measured. The measurements of these quantities can be carried out only if the signal samples are sufficiently denser than the signal wavelength, to allow the detection of the wave travelling from the ultrasonic sending transducer to the receiving one.

The longitudinal velocity, V_L, in the pulse-echo mode is defined as the ratio between two times the sample thickness, d, and time of flight, t, of the ultrasonic wave through the sample:

$$V_L = 2d/t \qquad (1)$$

The attenuation or attenuation coefficient [11], α, is defined from the decrease of amplitude, A, of a plane wave across the thickness:

$$A = A_0 \, exp(-\alpha d) \tag{2}$$

where A_0 is the amplitude of the incident wave. The coefficient α is measured in dB/mm according to:

$$\alpha = (1/d)\, 20 \log (A_0/A) \tag{3}$$

or in Neper/mm = 8.686 dB/mm according to:

$$\alpha = (1/d) \ln (A_0/A) \tag{4}$$

From measurements of acoustic velocity and attenuation, the components of a complex longitudinal or shear modulus can be calculated if a longitudinal or shear wave propagates through the sample, respectively. The propagation of shear waves is restricted by the very high damping properties of polymers often leading to a complete attenuation of the wave through the sample. On the other hand, the propagation of longitudinal elastic waves is easily observed in polymers as well as in reactive mixtures of monomers. When the sample dimension normal to the acoustic wavefront propagation direction is large, compared to the signal wavelength, the wave propagation is governed by the complex bulk longitudinal modulus L^* given by:

$$L^* = L' + iL'' \tag{5}$$

The storage (L') and loss (L'') bulk longitudinal moduli are related to the bulk (K' and K'') and shear (G' and G'') moduli [12]:

$$L' = K' + 4/3G' \tag{6}$$

$$L'' = K'' + 4/3G'' \tag{7}$$

L' corresponds to the stiffness of a system that is deformed changing its dimensions in one direction while in the other two directions the dimensions are constrained to be constant, as it occurs in samples where two dimensions are much larger than the third. For a liquid or a soft rubbery material $K' \gg G'$ and L' may be considered equal to the bulk modulus, while K'' and G'' may be of the same order of magnitude and the effect of G'' on L'' cannot be neglected.

The measurements of ultrasonic longitudinal velocity (V_L) and attenuation (α) may be used for the calculation of the storage (L') and loss (L'') bulk longitudinal moduli [12]:

$$L' = \frac{\rho V_L^2 \left[1 - \left(\frac{\alpha\lambda}{2\pi}\right)^2\right]}{\left[1 + \left(\frac{\alpha\lambda}{2\pi}\right)^2\right]^2} \tag{8}$$

$$L'' = \frac{2\rho V_L^{2}\left(\frac{\alpha\lambda}{2\pi}\right)}{\left[1+\left(\frac{\alpha\lambda}{2\pi}\right)^{2}\right]^{2}} \tag{9}$$

where ρ is the material density and λ is the wavelength, given by $\lambda = V_L/\nu$ as a function of the signal frequency, ν. These expressions may be also applied to the calculation of the shear or Young moduli when a different type of wave or sample geometry is used [12]. For low damping solids, like metals or ceramics, the attenuation may be neglected and a single longitudinal modulus can be calculated:

$$L = \rho V_L^2 \tag{10}$$

$$G = \rho V_S^2 \tag{11}$$

2.2. Ultrasonic Systems and Methods of Detection

The ultrasonic data acquisition system comprises at least a wave pulsing/receiving transducer or a couple of transducers, and a PC card inserted in the mainboard of a computer, which consists of all the components involved in ultrasonic equipment like a pulse generator operating in a range of frequencies at a certain repetition rate, an oscilloscope, a wave filter and amplifier, a A/D converter and a data processor. The set up can be computer-controlled through graphical software for automated measurement acquisition.

The train of electric pulses generated by the wave generator is transmitted into the material as incident microsismic waves by a piezoelectric transducer, in contact with the specimen or the structural element to analyse through a coupling agent (water, glycerine, honey, vaseline). The ultrasonic waves propagate into the material volume and are detected from a receiving transducer positioned on the opposite face of the element (signal transmission mode) (Figure 1).

In a different operating mode, the signal wavefronts can be reflected at the rear surface and detected at the front surface by means of the same transducer alternatively working as transmitter and receiver (pulse echo technique). The recorded echo signals contain several oscillations as the ultrasonic signal is reflected a number of times between front and rear surfaces and other anomalies encountered along the wave path. However, the amplitudes of the echo signal are gradually decreasing with propagation time due to attenuation. At each measurement, the pair of transmitted and received waveforms are evaluated so that the time of flight of the signal and its amplitude can be measured. From these, the ultrasonic signal propagation velocity in the specific material and the signal attenuation can be calculated, respectively, and consideration can be drawn on the health state of the specimen [14].

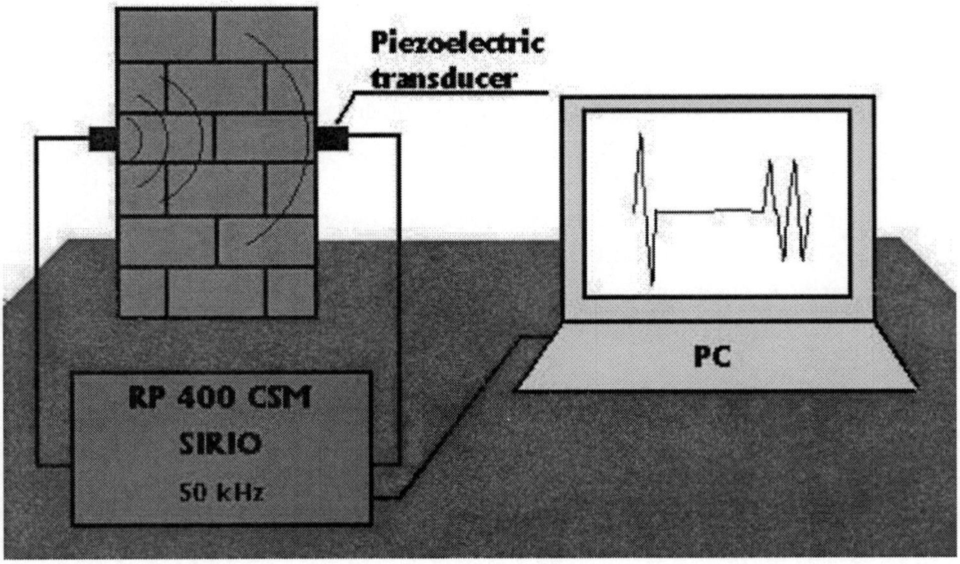

Figure 1. Ultrasonic system.

2.3. Sonic Technique

The physical principles of the sonic technique are very similar to those of the ultrasonics ones since both techniques belong to acoustic methodologies. With regards to operating frequencies, in the sonic field, signal frequencies are restricted below about 20 kHz. This means larger wavelengths and lower resolution than ultrasonics but with the advantage of reaching greater penetration depths into the material. In fact, there is a trade off between frequency of signal employed, resolution and penetration depth obtained.

The pulse function can be approximated to a sinus curve, whereas the width, and corresponding contact time of the signal source produced by impact with the surface, depends on different parameters, such as the stiffness and diameter of the used impact mass (commonly, an instrumented hammer) or the stiffness of the structure surface. As a result of a larger impact/surface contact area (i.e. in case of low stiffness or large sphere diameter), the contact time increases. Therefore, the generated wave will be characterised by lower maximum frequencies and a narrower frequency range.

The use of higher frequency signals would be preferable because these have shorter wavelengths, thus providing better resolution, i.e. capacity to detect small and/or superficial defects. During signal propagation, higher frequency components are more likely to be quickly attenuated and lost, mainly due to geometric attenuation, as the wave front widens from the source point on the testing surface into the volume of material, and due to structural attenuation because of friction and energy loss during material particle displacements at propagation of elastic waves, and scattering from inhomogeneities in the material (such as aggregate grains in concrete, porosity, microcracking or other boundaries like mortar joints in masonry). When anisotropic, heavily damaged or thick materials have to be investigated, in practice it is more convenient to adopt the sonic technique instead of ultrasonics.

2.4. Sonic System and Method of Detection

The sonic instrumentation is easy to use both in the lab and on site. It differs from ultrasonic because it makes use of a modally tuned instrumented hammer as signal transmitting device, whilst the receiving probe is commonly a displacement transducer or an accelerometer. The generation of the signal via an elasto-mechanical impact on the material surface through the use of the hammer offers the advantage of producing a more energetic signal, thus less prone to be rapidly attenuated. The propagation of impact generated stress waves in a heterogeneous bounded solid such as a concrete or masonry element is complex. Briefly, *R*-waves travel along the surface spreading radially outwards, whilst *P* and *S* wavefronts travel spherically in the semi continuous materials (Figure 2), until inhomogeneities along the path modify intensity of propagation in terms of velocity and attenuation. Due to higher energy and velocity of compressive *P* waves, the sonic technique commonly relies on the propagation and reflection of these [15].

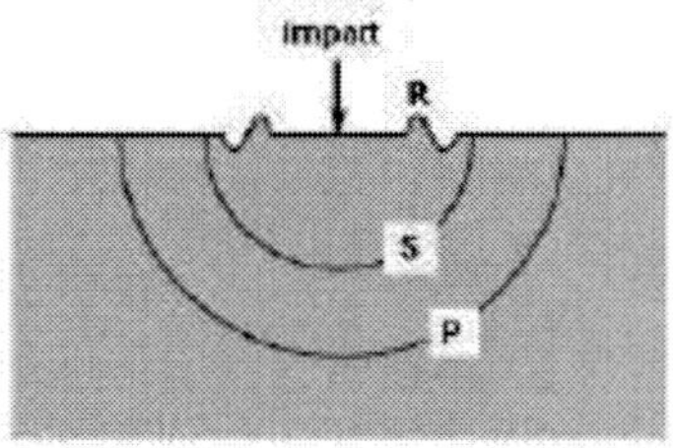

Figure 2. Schematic of elastic impact on the material surface producing Rayleigh waves *R* on the surface, shear waves *S* and compressive waves *P* into the solid.

In sonics, various methodologies of data acquisition can be differentiated depending on relative location of transmitter (hammer) and receiver on the element. In direct through transmission of the wave, transmitter and receiver are located in front of each other, on opposite sides of the element (Figure 3a); in semi direct transmission, transmitter and receiver are not located in front of each other, on opposite sides of the element or on not parallel surfaces (Figure 3b). The combination of these 2 operating modes is also proper of sonic tomography data acquisition, as shown later. When transceivers are located on the same surface, a certain distance away from each other, indirect or superficial measurements are collected (Figure 3c). Finally, the reflection mode foresees transmitter and receiver adjacent to each other (Figure 3d). These last two modes are adopted when only one surface of the tested element is accessible.

Before starting data acquisition, it is good practice to verify the typical frequencies and propagation characteristics of the signal obtained by the hammer impact on a representative area of the specific material, and to evaluate if these are appropriate to the aims of testing. Two operators generally occur for data acquisition: one at hammer, the other operating receiver and data acquisition system.

At each measurement path, a couple of signal waveforms are recorded: the first is the impulse generated by the hammer at the impact, the second is the signal received after propagation in the material, along the path (Figure 4 left). Each measurement reflects the average characteristics of the material along the path. Data analysis consists of measuring the signal propagation time and/or attenuation from the recorded waveforms (Figure 4 right). The accuracy of signal velocity determination depends on the exact measurements of the path length (precision ought be within ±1%) and of the time of flight. A number of parameters may affect the measurements, including the rughness of material surface, moisture content and temperature of material tested, path propagation length, geometry of the element, stress state, presence of metal inclusions, and, of course, cracks, voids and defects.

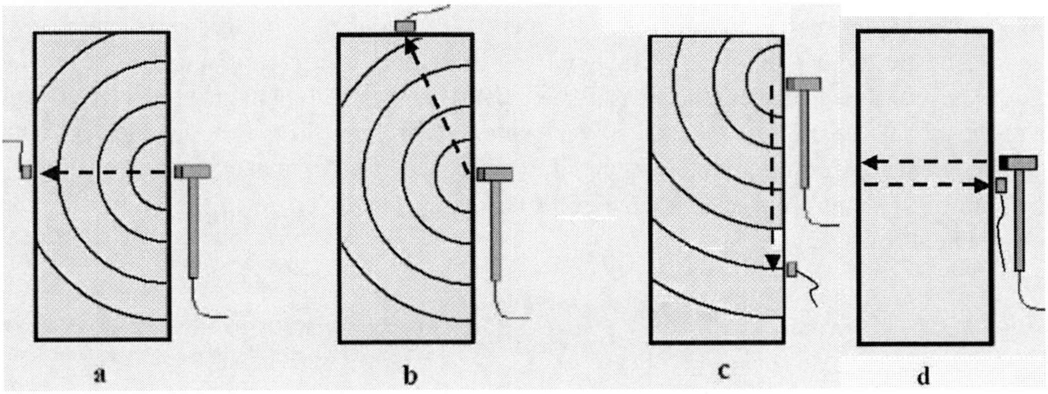

Figure 3. Transmission modes for sonic wave tests: a) direct, b) semidirect, c) indirect. d) Reflection mode.

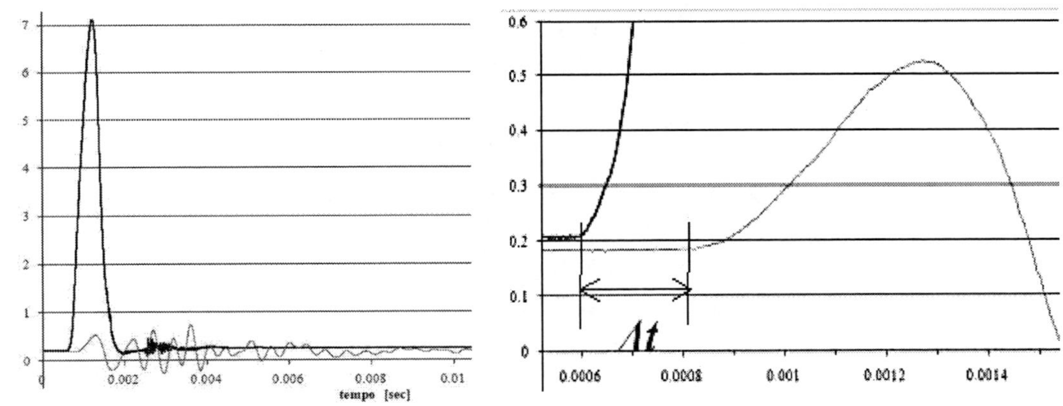

Figure 4. Example of a measurement: hammer waveform in black, received signal in grey (left). Detail of the beginning part of the 2 signals for time of flight (Δt) analysis (right side).

3. Innovative Techniques

3.1. Tomography Analysis

Tomography, from the Greek root "tomos", meaning slice, involves the iterative reconstruction of a section of the test specimen using a high number of measurements made from station points on its perimeter along paths through the specimen. The outcome of data inversion is a 2-dimensional image of the section, mapping the measured material parameter and representing the condition state of the tested element. The mathematical theory was established by Radon in 1917 in which it is shown that the internal characteristics of an object can be exactly reconstructed by a complete set of projections through the object. This theory has been used in geophysics and extensively in modern medicine. More recently, in the structural field it has had an encouraging use for the assessment of concrete and masonry [17-19] as well as timber structural elements [20]. Tomographic reconstruction can be applied to acoustic techniques such as ultrasonics and sonics [21-23] applied to bridges and dams.

In acoustic tomography, for each signal source station, a discrete number of receiving stations are considered, so that the measurement paths can be ideally imagined as straight rays of a sort of fan departing from each source station. Therefore, these rays are incident with varying orientation on discontinuities possibly contained into the plane section considered. The sum of ray fans from each source point makes up the coverage of the section by the measurement paths (Figure 5).

Figure 5. Example of the section ray path coverage for transducer arrangement on 4 sides (ideal straight tomographic paths).

The basic concept is that the stress pulse on each projection travels through the specimen and interacts with its internal construction. For each receiving position, the projection will be affected by the physico-mechanical characteristics of the crossed material so that variations in the internal conditions result in different times-of-flight being measured. The tomographic software reconstructs the section by combining the information contained in a series of these projections, obtained at different angles through the section [16]. For example, the signal time of flight recorded between 2 stations will reflect the average velocity along the path connecting a particular transmitter with the receiver considered or, better, each time of flight represents the integer of the inverse of speed along the path. Acoustical signals have the tendency to travel faster in denser materials, so that velocity will be higher in sound material and real measurement paths will not follow straight lines. When encountering a void, the wave front will go around the cavity, lengthening its path and causing an apparent lower velocity to be measured.

In order to reconstruct the velocity distribution in the section, it is possible to use the technique of expansion in series, which considers the continuous function of the inverse of velocities as a series of discrete elements, called pixel, or areas in which velocity can be considered to be constant. Known all the time of flights and the section geometry, it is possible to calculate the solution of the tomographic problem thanks, for example, to algebraic reconstruction techniques or newer algorithms. During iterative reconstruction, an initial velocity model across the section is progressively modified up to convergence to the solution, thanks to correction factors calculated from residuals left from each calculated time of flight model.

The greater the number of measurements taken, the more accurate will be the tomography results but, due to time and cost of data acquisition and analysis, the number of recorded projections cannot be infinite. Thus, for each specific case, before starting data acquisition it is fundamental to design the orientation and number of tomographic measurement paths across the section. Once the section geometry is precisely known, and in function of signal wavelength and size of discontinuities to be sought, it is necessary to consider various hypotheses of station distance along the perimeter, opening angle of the measurement fans and rays in the fan (Figure 6). This in order to obtain a section coverage possibly dense and homogeneous, and to avoid spurious effects in the produced plots [17]. This care is particularly important when the section area is irregular or access is not possible from all perimeter sides: often distribution of measurement points has to be limited to 3 or 2 sides.

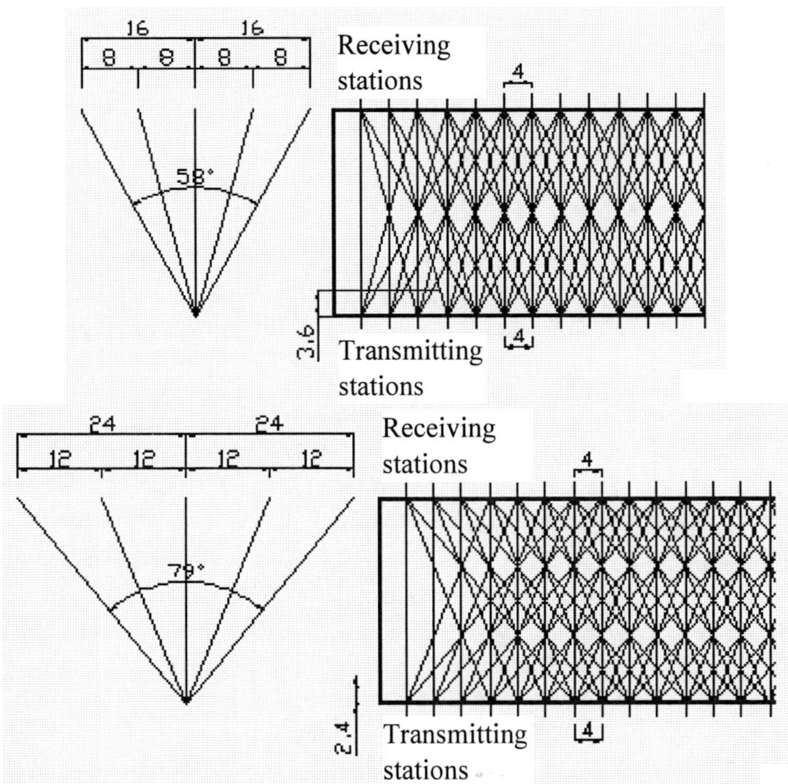

Figure 6. Two example hypotheses of area coverage obtained by different angles of ideal straight measurement rays and same distance between stations, in 2-side tomography. On the left side of each

section, a single fan of measurements is shown, each made up of 5 rays. As it can be noted, each fan is a combination of one direct through transmission measurement and 4 semi direct transmission measurements (see Figure 3a and b).

As an example, in Figure 7 upside, it can be seen how, by using the same number of transmitting stations (12 stations) but differently placed along the perimeter, in case "b" a better coverage can be obtained both in terms of homogeneous density and corner area coverage; in Figure 7 bottom, both drawings foresee fans made of 5 rays, but the smaller fan aperture of case "b" assures even coverage by combination with a higher number of stations.

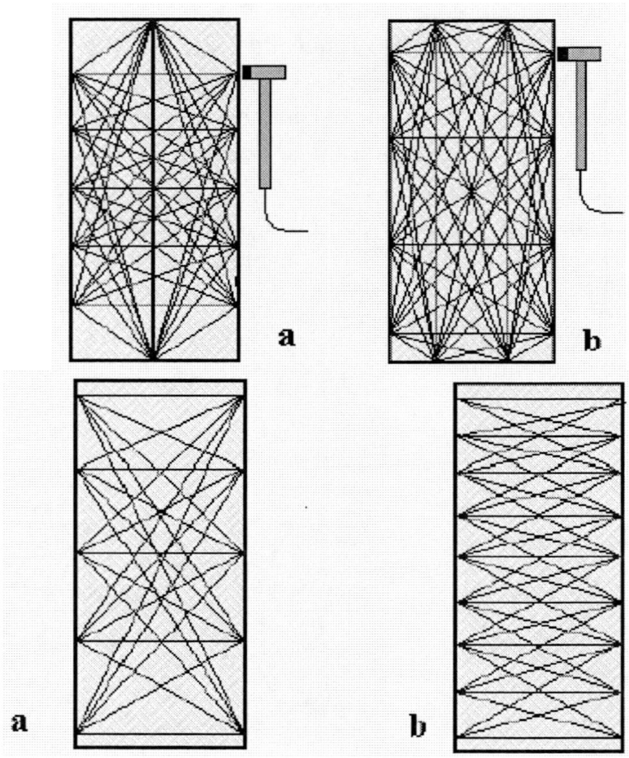

Figure 7. Examples of various station distribution and section coverage obtained (Colla, 2009).

3.2. Impact Echo

The Impact-Echo (IE) method has been used since the second half of the 80's for NDT of concrete structures. Its applications are multifaceted and, on concrete structural elements, they include the location of voids, faults, decay, delaminations as well as the thickness determination of single and multi-layer structural components [26-37]. Similar to sonic reflection measurements, the IE method is a wave propagation-based technique and consists in measuring the reflections of an acoustical wave (typically in the range from 0 to 60 kHz frequency) produced by a short elasto-mechanical point impact at the surface of the material and propagating into the element as compressive, shear and surface waves until reaching an interface (material impedance change). If the lateral dimensions of inhomogeneities, possibly present along the wave path, are not too small referred to the signal wavelength, the waves will be reflected and detected via a transducer arranged near the impact point on the surface, so that compressive wave arrivals will be dominant. A waveform is so built up in the time domain by the first and subsequent reflection arrivals. Nonetheless, IE differs from sonics in the fact that the hammer does not need to be instrumented (usually steel sphere of various size are used, Figure 8); only the received full waveforms are recorded and not the signal produced by the hammer, as not the time of flight, instead the multiple wave reflections between front and rear sides of the element are used in the analysis (resonances make up a time curve, which is a few milliseconds long); data analysis is carried out in the frequency domain after FFT has been performed on the time waveforms and it consists in locating and interpreting peaks of interest, their measured frequencies and signal patterns in the obtained spectra (graphs of amplitude versus frequency components).

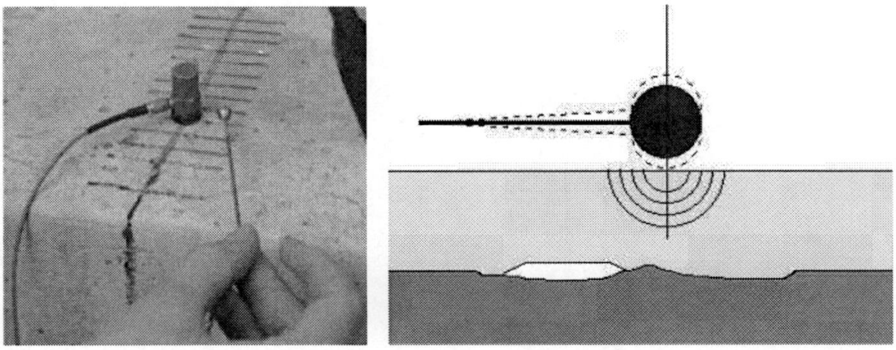

Figure 8. Wave generation in impact-echo.

Predominant frequencies due to multiple reflections of waves within the structure appear as peaks in the spectrum. An important evaluation criterion is the magnitude of the measured peaks. This parameter is strongly influenced by the range of frequencies excited by the impact, thus it is necessary to properly select the impact source in order to produce an optimal frequency range tailored to the test aim and so to ensure accurate data interpretation.

Waves are almost completely reflected if the second material is air, such as in the presence of a void or at the external boundaries of the element under investigation. By knowing or estimating the wave propagation velocity in the material, the corresponding element thickness and interfaces depth can be evaluated using the basic equation that relates wave velocity (v_c), thickness of the element (t) and measured frequency of wave reflection (f):

$$t = v_c /2f \quad (12)$$

This equation can only be used for reflections at interfaces with lower impedance than concrete (i.e. air). For materials with higher impedance, such as steel (i.e. in presence of reinforcement and tendons), a modified relation has to be applied, where t_{steel} is the concrete cover to the steel inlay [36]:

$$t_{steel} = v_c /4f \quad (13)$$

The IE technique is born as a point method giving often good and encouraging results: the operator, based upon his/her knowledge of the method and of the structure, or at random, selects a number of points for testing. With this method of local, punctual readings, important information can be missed when the points are badly selected or not enough. Especially in the presence of complex element geometry (box girder bridge, multiple ducts, crossing reinforcements), the experimental results are not always unequivocal and can easily lead to misinterpretation [30].

A scanning impact-echo implementation of the method has been later developed with the aim of easing data collection and interpretation [27]. An automated data collection procedure aimed at recording reliable data, independently from the operator and IE has become more widely employed in civil engineering. The proposed scanning procedure consists in collecting a complete set of data along measurement lines, at stations closely and regularly spaced, so to plot a cross sectional image representing the concrete element under investigation. This enables the user to easily identify features of interest in the data. By collecting data on an array of parallel lines, testing areas can also be covered. Frequency series and impactechograms are 2-dimensional plots of data, which present on the horizontal axis the position of the reading station and on the vertical axis frequency (Figure 9).

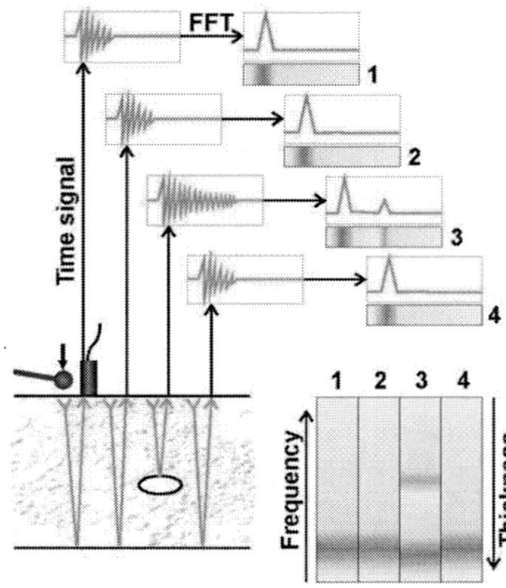

Figure 9. Schematic of the scanning impact-echo procedure.

Different grey or colour tones represent values of the frequency amplitude. If the velocity of wave propagation in the element is known, slices of position vs. depth can be plotted.

More recently, other variations of the method have been implemented, such as SIBIE, SAFT and Air-coupledIE.

3.3. Monitoring System

Non destructive testing by means of ultrasonic waves can be used not only to verify in situ the structural integrity of concrete in civil engineering structures, but also to monitor modification of material properties in time. Development of efficient methodologies to determine the presence, location, and severity of hidden damage in critical structural components is an important task in the design and construction of structural health monitoring systems in aging as well as new structures [37]. For example, after casting, cementitious materials exhibit a continuous change in the mechanical properties with time due to a chemical reaction known as hydration. In addition, these materials also exhibit a change in state, from a fluid to a solid, in the first few hours

Conventionally, changes in the material properties are obtained by monitoring the visco-elastic moduli over time before setting, and the increase in the elastic material properties of the solid cementitious material, after setting. An ultrasonic test setup and the data analysis procedures, which provide for continuous monitoring of the hydrating cementitious materials from a very early age through setting are presented in [38]. Experimental evidence suggests that initially there is a larger relative increase in the viscous component of the shear modulus while later there is a larger relative increase in the elastic component.

For this reason a testing device has been developed which utilizes the velocity of ultrasonic waves in order to continuously monitor the setting and hardening of cementitious materials. The method can be used for quality control in production of admixtures and new binders and for control of constancy of concrete production [39].

To avoid the problem due to the indirect ultrasonic pulse velocity method (the application of this technique is rather time consuming (15 min/test) and requires an almost flat surface, which makes it generally not appropriate for shotcrete or if there has been spalling), it was proposed another procedure allowing rapid interpretation of the results with no need for a preliminary calibration for the specific material properties: the embedded piezoelectric transducers for cement paste hydration monitoring. Since the embedded transducers had good coupling with the cement paste, reliable test results have been obtained. It is feasible to use the technique as an in situ technique to monitor the hydration process at an early age, as well as the health condition after maturing for a concrete structure [40]. However, the possible influence of cracks, delaminations and presence of distinct layers could markedly affect the results and a careful check on both their repeatability and consistency is recommended, regardless of the interpretation procedure.

Furthermore, with ultrasonic testing methodology it is possible to monitor the behaviour of Fiber Reinforced Plastic (FRP) for the assessing of the structural integrity of FRP wrapped concrete [41].

FRP composites or advanced composite materials are very attractive to use in civil engineering applications, due to their high strength-to-weight and stiffness-to-weight ratios, corrosion resistance, light weight and potentially high durability. Their application is of most importance in the renewal of constructed facilities infrastructure such as buildings, bridges, pipelines, etc [43]. Recently, their use has increased in the rehabilitation of concrete structures, mainly due to their tailorable performance characteristics, ease of application and low life cycle costs. These characteristics and the success of structural rehabilitation measures have led to the development of new lightweight structural concepts utilizing all FRP systems or new FRP/concrete composite systems as reported in the overview of the research and development of applications of advanced composites to civil infrastructure [44]. The planning and the application of composite materials in civil engineering does not leave out of consideration about the limits of the " reinforced system in work" [45]. The state of art of the research on the composite durability under different environmental conditions shows that a lot of work had been done on the behaviour of the single constituent of the composite, but few researches have been done on the reinforced system in service.

The European standard (FIB – Federation International du Beton) does not introduce any quantitative request, giving only information about the possible problems, while the USA standard (ACI – American Concrete Institute) introduces the coefficients of reduction of the resistance of the FRP due to the aging of the composites. The problem of the aging of structural elements reinforced with FRP is qualitatively defined, it is necessary now to find quantitative characterization in situ that can optimize the existing solutions and allow the development of new approaches.

4. In Situ Applications: Case Studies

4.1. Investigation of the Mechanical Properties of Concrete By Means of Ultrasonic Techniques: Punta Perotti in Bari (Italy)

Concrete is a material with a very heterogeneous composition. This heterogeneousness is linked up both to the nature of its constituents (cement, sand, aggregates, reinforcements, protection,.) and their dimensions, geometry or/and distribution. It is thus highly possible that defects and damaging should exist. Non Destructive Testing and Evaluation on this material can aid to detect the changing in the mechanical behaviour during the time.

To explore this theme a systematic study was carried out [46] on a building (Punta Perotti) in Bari (Italy) before its demolition. Punta Perotti building (14 levels high, about 1560 m^2 for level) was exposed for more than 10 years to particular aggressive environmental conditions (near the sea and in a windy area), of which are well known the technical schedules of original materials, was selected in order to study the life cycle of the materials and structures, until the demolition process. About 180 cores (10 cm diameter) and 290 microcores (2 cm diameter) were taken from the pillars of the structure made in concrete at different floors (-1, 0,1,2,3) after the non destructive testing (ultrasonic tests and rebound hammer tests) made on the same pillars. The European [8] standard and USA [9] guide lines were followed both in the in situ measurements and in the data analysis.

The main object of this work was to test statistically a new method to identify a good practice to select the eligible test stations of the same structural element on which to do the cores to minimize the errors in the technical evaluation. In fact although NDT are largely regulated, such control is still relatively low because it is necessary that the staff is well trained to perform such tests, because there are some strong points to the discretion of the technician (as definition of the measuring points to make the analysis). We decide to work with ten cores (decuple) because it is the number of the cores indicated by the Italian Guide Lines [47] for a structure as Punta Perotti to obtain an adequate knowledge of the building for each floor.

In order to verify the hypotheses of pre-selected zones on which carrying out the cores, an exhaustive analysis of all the possible decuples of cores was carried out. The number of possible decuples for each floor is reported in table 1.

As it is shown in Figure 10, the curves built up for the second floor on the assessment of European standard and America guide line are an underestimation of the ‘ad hoc correlation curve’.

Our main goal was to try to be closer as possible to the ad hoc correlation regression of the entire population with the identification of decuples more significant through the use of non-destructive tests carried out previously.

A new method was studied to identify the structural element from which extracting the cores for the mechanical testing. In our opinion it can be useful for the engineers to obtain and elaborate data that have a higher probability to be representative of the correlation curve of all floor.

Table 1. Number of the possible decuples for each floor in function of the available cores for each floor

Flo or	Numbe r of cores	Number of cored pillars	Number of decuples
-1	47	18	269.034.63 1
0	14	10	16
1	18	10	256
2	56	20	3.348.291.3 35
3	19	11	1.792

It was, so, supposed to take into account the decuples with a distribution of the ultrasonic velocity measurements similar to the one calculated on all the data of the floor in order to find a new method to identify the structural elements from which collect the cores for the mechanical tests and obtain data with a higher probability of being representative of the correlation curve of the floor (Fig 11-12).

The number of decuples available it was so reduced and then was measured the number of decuple included in various intervals (from -5 to +5) around the representative correlation curve of all samples related to the floor 2.

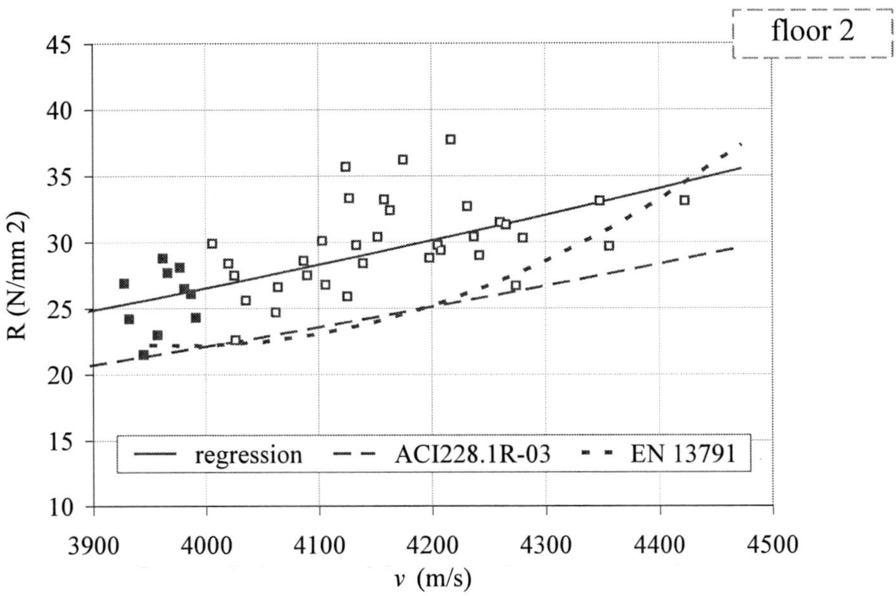

Figure 10. Comparison between the ad hoc correlation curve, UNI-EN 13791:2008 standard and ACI 228.1R-03 guide line obtained for all the cores of the 2nd floor .

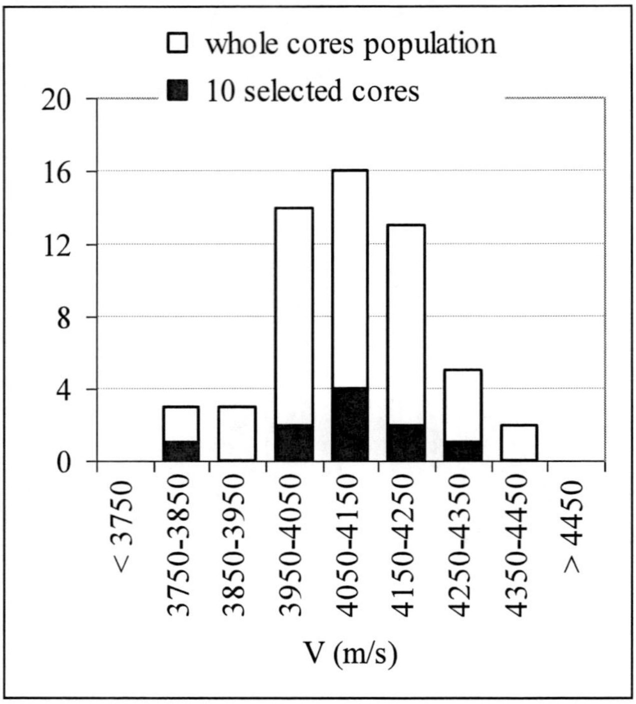

Figure 11. Distribution of the ultrasonic velocity measurements of the 10 selected data considering similar velocity statistical distribution.

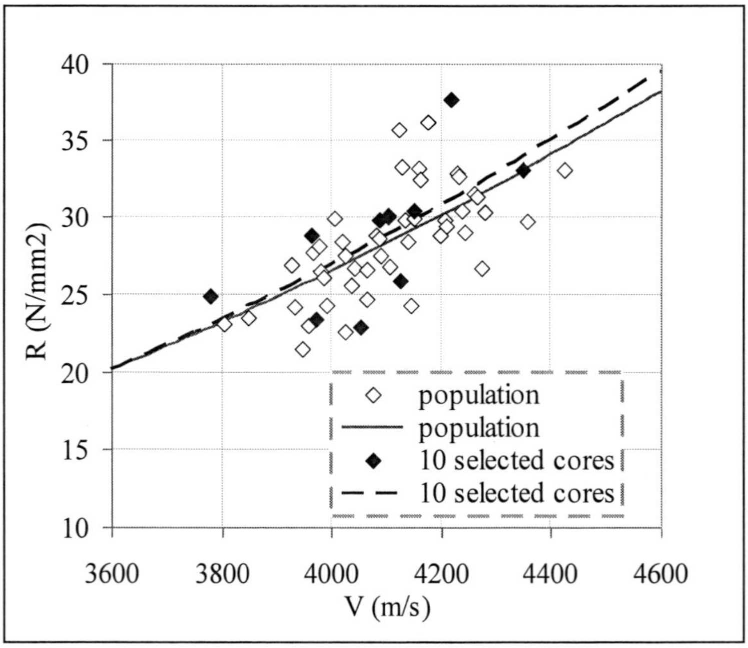

Figures 12. Distribution of the regression curves of the 10 selected data considering similar velocity statistical distribution.

The results of the analysis demonstrate that with a previous campaign of non destructive testing and an accurate statistical analysis of the results, it is possible to improve of 3 % the probability to obtain correlation curves representative with the selected decuples.

Especially following this method and analyzing the data using the technique of the statistical similarity of the distribution of the ultrasonic velocity it can be reduced the wrong choices of the selection of the decuple that lead to the definition of correlation curves absolutely not representative of the state of the concrete structure under consideration.

4.2. Sonic Technique Assessment of Crack Pattern in Masonry

Where possible, sonic tests are carried out in direct through or semidirect transmission modes (see Figure 3). Nonetheless, when only one element surface is accessible or free from plaster (the presence of a plaster layer would affect significantly the signal propagation and, usually, the operative procedure requires removal of the plaster before data acquisition. In the case of historic constructions, where the finishing layer cannot be destroyed and replaced, this can constitute a strong limitation to acquiring good quality signals) data can be acquired in indirect transmission mode.

Data are then commonly collected at reading stations placed at the nodes of a grid with regular squared or rectangular mesh. In the following, an innovative example is presented of superficial measurements on a damaged brick masonry panel. The tests, aimed at investigating the crack pattern present in the wall and at proposing a data interpretation based on visual extraction of information from 2-dimensional data images (image diagnostics), were carried out at stations distributed on concentric circles. The procedure using this type of grid has been tested on site on stone masonry archaeological structures in order to evaluate the quality of different masonry layouts, their homogeneity of behaviour and conservation state [48]. This operative procedure has the advantage, in addition to maintaining static the transmitter (or the receiver) thus collecting readings on progressively larger masonry areas, to consider additional directions of measurement paths with respect to traditional vertical and horizontal directions, and so to catch anomalies with various orientation and obtain a richer information about the masonry area undergoing testing. This is particularly relevant for the case of historic structures with highly inhomogeneous masonry.

The brick masonry panel non destructively tested in the lab had previously been reinforced with a GFRP tissue on one surface and it had undergone static mechanical testing by in-plane compression and shear load up to formation of extended cracks. It has overall dimensions of 1,27 m base × 1,25 m height and 0,128 m in thickness and is made of 17 courses of full bricks. The passing-through crack pattern, visible only on one surface of the panel, due to the GFRP reinforcement, presents a main crack in diagonal direction following the mortar joints (Figure 13 top).

The sonic data have been collected both along path crossing and not crossing the main crack. For example, one of the grids used (Figure 13 bottom) has foreseen 42 receiving stations on 5 concentric circles respectively with ray length of 10, 20, 30, 40, 50 cm, along rays every 45°, whilst the transmitter remained in the centre. Receiving stations have been numbered clockwise starting from above and from the smallest circle. The acquisition software has allowed visualizing data in real time both as single waveforms and 2-D images of the received signals.

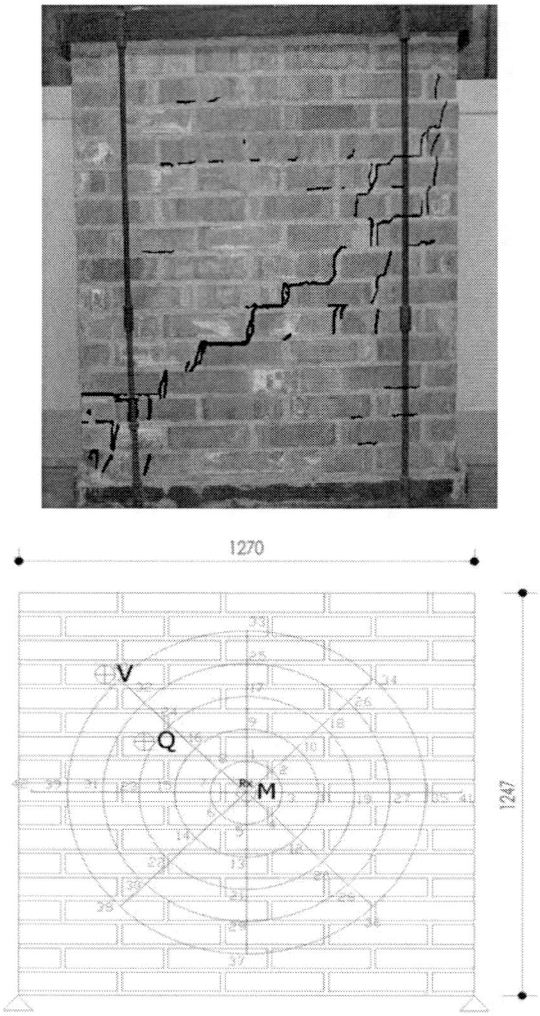

Figure 13. Masonry panel with crack pattern (top) and grid of reading stations with transmitter static in position *M* (bottom).

In Figure 14, the 2D image visualizes in grey tones the 42 received row waveforms (starting from below in the image) [49]. On the x-axis, the time interval goes from 0 to 4 ms, whilst on the vertical axis it is reported the number of registered waveforms. The grey shades represent the amplitude of oscillations in the single waveforms according to the colour legend shown in the top right corner of the image and which has in black the maximum positive amplitudes and in white the maximum negative amplitudes. On the time axis, the shortest signal "time of arrival" are those from the first readings - with shorter measurement paths -, whilst the times of arrival progressively increase with increasing path length. This is in accordance with expectations for a homogeneous, isotropic, integer material, but this is not our case.

Indeed it is noticeable that beyond the first 12-14 readings, a rhythmic and repetitive ongoing is observable both with regard to times of arrival and their amplitudes. The variations of these 2 parameters can be put into relation with the direction of the testing path, its length and number of mortar/brick interfaces crossed. Furthermore, a strong influence exists from the crack pattern.

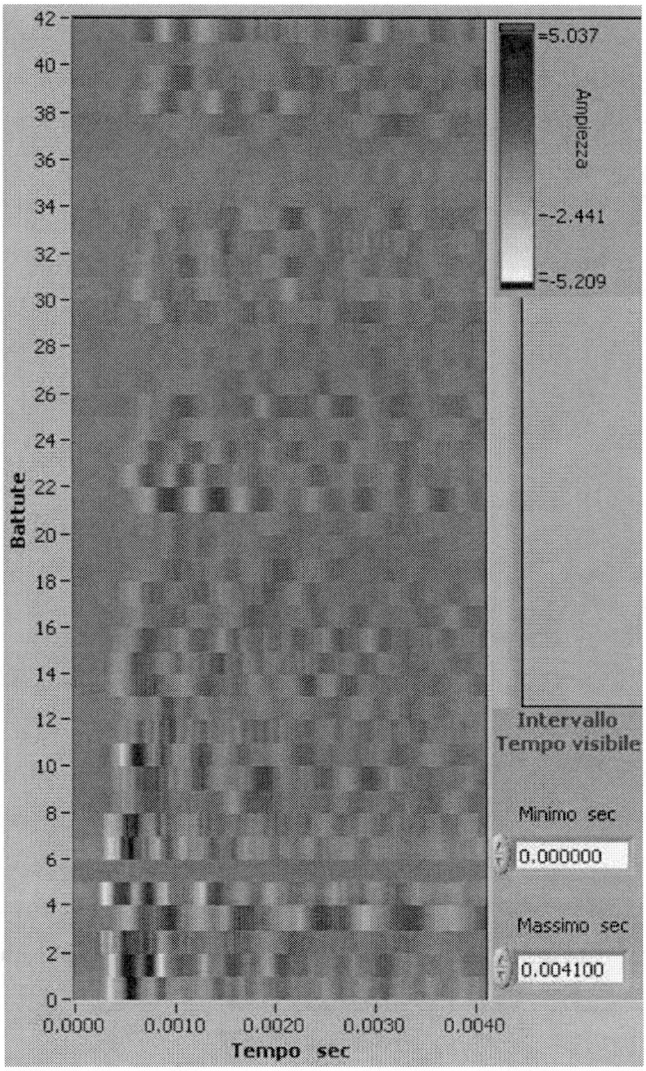

Figure 14. 2D sonigram of 42 readings from stations depicted in grid of Figure 13 bottom.

Data at stations 7, 15, 23, 31, 39 and 42 show the greatest signal amplitudes and shortest arrival times in the respective circles. They correspond to stations located on the horizontal direction at the left of the transmitting station, that is on a masonry course visually integer and with only 2 mortar joints. Instead, the relative minima of signal amplitude and relative maxima arrival times are recorded at stations 4, 12, 20, 28 and 36, that is along the diagonal direction perpendicular to the main crack. A different observation can be made for the first 8 readings, recorded at the smallest circle. These are expression of local values of travel times and amplitudes, that is they are mainly affected by the single materials crossed (mortar, brick) more than by signal propagation in the masonry composite. This discrimination between data, carried out by visual observations, is important because the threshold between local and global measurements is affected by masonry units' dimensions and, influencing the calculating signal velocities, alters the average values, with consequences on the evaluation of the masonry quality. Such observations are even more important in the case of testing on masonry having units highly inhomogeneous in dimensions.

All above considerations have been made possible by visualising the data as 2-D images and can be carried out on site, if necessary, with evident advantages on data analysis costs and easiness of interpretation. In turn, this has positive repercussions on the diagnosis and assessment of the structural element. Thanks to qualitative observations regarding propagation speed of single sonic signals and their attenuation, it has been possible to relate data anomalies with the position of reading stations on the masonry surface, and locate in a quantitative manner the extension of the damaged area.

4.3 Application of Sonic Tomography Analysis: The Case of a Historic Masonry Bridge Abutment

Masonry arch bridges form a large proportion of the bridge stock all over the world and this will continue to be the case for a long time to come, since the vast majority of these bridges are still in perfectly serviceable condition. Periodic assessment of their load carrying capacity will remain an essential part of the bridge authorities' management activities for the foreseeable future. The outcome of these assessments can be very significant in terms of resulting financial and logistical burdens for the authorities. It is therefore in their interest to seek, wherever possible, improvements in the methods of assessment.

Bridge authorities have sponsored major research projects in recent years to develop numerical methods for predicting the collapse strength of masonry arch bridges. A number of newly developed methods were then tested against the results obtained from full-scale tests [50]. The overall lessons learnt from the developments so far can be summarised as follows.

Idealistically, the structural system of an arch bridge, composed of a number of elements: i.e. the foundations, the barrel and the surrounding fill, can be conveniently analysed using advanced numerical methods of analysis, such as the finite element method. However, in practice, the accuracy of the results becomes questionable due to a multitude of complexities (including that most arch bridges do not have reliable records of construction or repair details, and it is difficult to know the physical dimensions of the main structural elements, or the presence or otherwise of additional features such as the haunching at supports, saddling over the barrel or internal spandrel walls and ribs) that make the computational precision of the analytical assessment methods largely become immaterial. Thus, the results from these programs cannot be relied upon to give a black and white indication of the capacity of a bridge, since these analyses are dependent upon the input data.

Experience has shown that results from any calculation based arch assessment method may have to be viewed as giving only a qualitative assessment of capacity, and in order to decide on the appropriate action, one may have to rely on additional, perhaps qualitative, information obtained by other means. Thus, if an arch bridge fails assessment by a small margin, and it is not showing significant deterioration, and it is suspected that additional structural features are present within, it may be reasonable to consider the bridge to be satisfactory for the time being and perform NDT techniques that may provide a relatively quick and inexpensive means for making a qualitative judgement on the likelihood of a bridge having any additional reserves of strength.

Non-destructive investigation surveys by means of radar, conductivity and sonic tomography were undertaken on one abutment of a historical stone masonry structure, which was known to present the problems described above [52] (Figure 15).

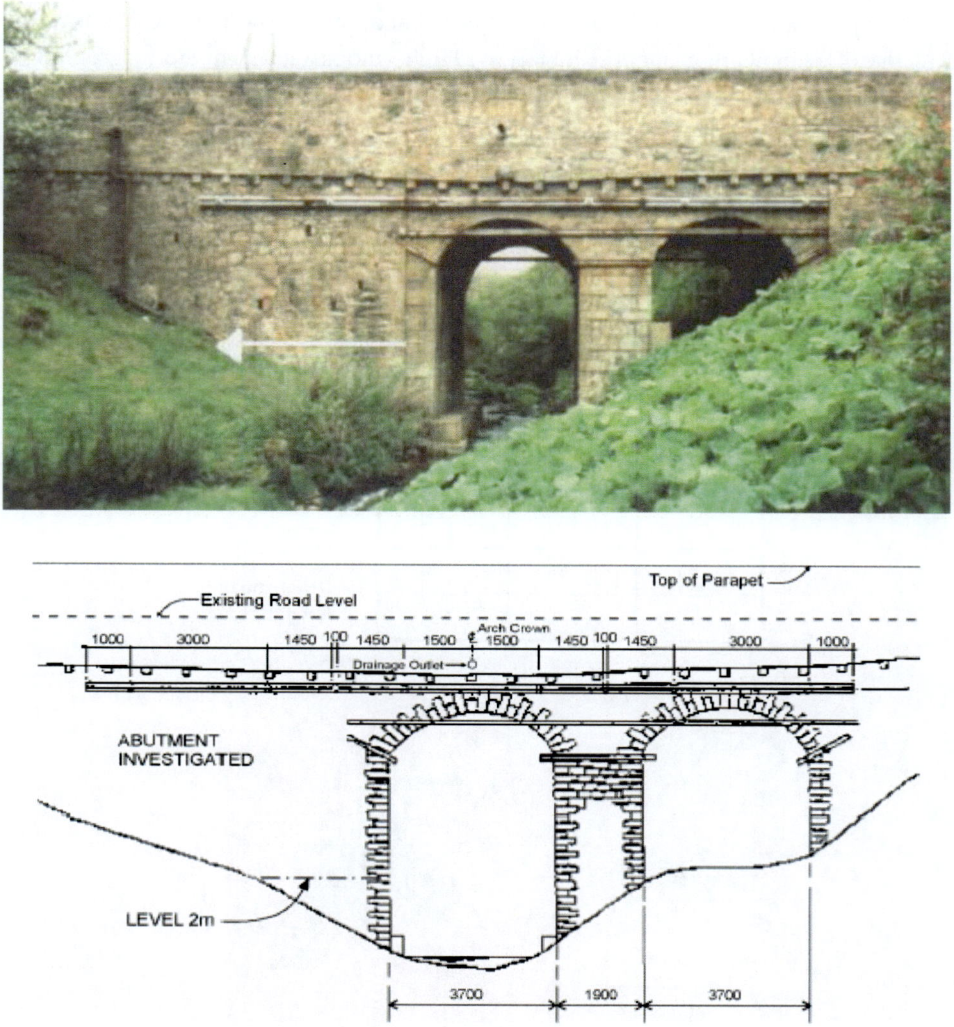

Figure 15. Elevation of North Middleton Bridge, Scotland.

Time domain impulse sonics were undertaken by measuring the time taken by an impulse from a modally tuned instrumented hammer of adequate size to transmit through the structure. On a horizontal cross section, measurement stations were marked, one meter apart, on the upstream, downstream side and below arch face of the abutment. Due to different slope of river embankment, the number of stations was different on upstream and downstream side. Each station was alternatively used as wave source point, whilst the wave was picked up by the receivers in all stations, located on the other 2 abutment faces.

Sonic velocity tomographic plots were constructed using the first arriving wave energy transmitted through the abutment after interpreting the travel time of single ray paths (Fig 16 bottom). An image is built up by an iterative process, mapping sonic velocities distribution through the various materials in the cross-section. The sonic velocities are related to material densities. The cross-sectional representation is an over-simplification as the ray paths will not be in straight lines in reality.

A sonic tomographic model was constructed, using first the simplifying assumption of straight ray paths and sub sequentially refined by using bending ray paths to the model

(Figure 16 top). Values of wave velocity between medium to low are located towards the centre of the plot, indicating backfill materials. High velocity areas relate to grouted zones. By applying constraints to the model and re-iterating, masonry thickness on the downstream side was comparable with the one determined from the conductivity survey undertaken.

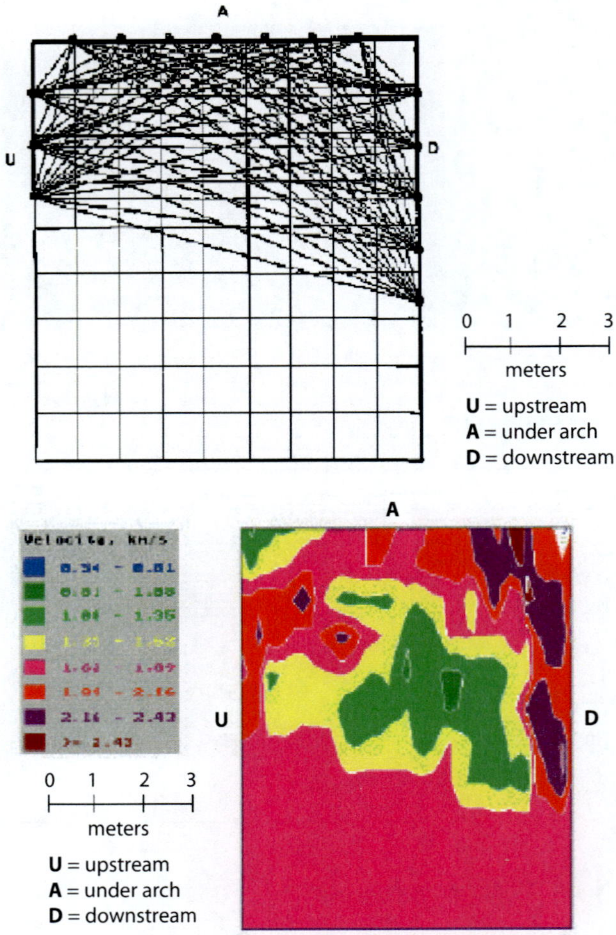

Figure 16. Sonic ray paths: initial hypothesis of ideal straight paths in tomography reconstruction (top). Tomography distribution of sonic velocity through the bridge abutment with constrained data (bottom).

Subsequent endoscopic investigation into the structure has confirmed the hypothesis and revealed that hollow compartments were built into the bridge and that some of them have been partially or totally filled by grout injection during earlier strengthening and reinforcing operations.

The NDT methods used in the project have demonstrated the ability to be rapid, low cost, techniques from which tomographic cross-sections, inhomogeneity identification and layering within the masonry can be detected.

4.4. Impact-Echo Assessment of Grouting Condition in Post-Tensioned Concrete

Concern exists worldwide over the safety and durability of post-tensioned structures as a result of voids around the tendons in the ducts. Faults in tendon duct grouting permit the formation of voids in the duct, which can give rise to ingress of water and chlorides, corrosion of tendons, and increased stress. Inspection techniques, including available NDT (X-ray, ultrasonic, electrical resistance, magnetic perturbation, georadar, etc.), present a number of limitations and require engineering expertise and interpretation. Today's civil engineering needs are for fast-to-apply and easy-to-interpret NDT methods. Furthermore, a condition for the successful use of their outcome findings in the damage analysis and maintenance planning of the bridge stock is their capability of providing assessment information in quantitative terms.

The objectives of the work undertaken via a single NDT technique were the ducts lateral and depth location in the concrete beam, and the evaluation of the post-tensioning condition by identifying possible voiding in the metal ducts. The scanning impact-echo method was applied on site on part of a 30-year old pre-cast concrete I-beam which was made available for field measurements in the frame of the dismantle phases of a highway bridge.

Thanks to the 2-dimensional type of data collection and imaging, it was possible to visualise in real time cross-sections and depth slices of the concrete beam web containing the metal ducts (67 mm in diameter) and to accurately locate the post-tensioning ducts. Signal interpretation has allowed recognizing signal patterns in the impactechograms produced and identifying voiding in the ducts discriminating reliably between fully grouted and partially grouted ducts. The NDT data interpretation was compared with the condition of the ducts as visible from large cores (200 mm in diameter) sampled from the tested areas, and satisfactorily verified.

The impact-echo measurements covered circa an area of 1 m by 1 m on the girder's web. The area coverage was obtained by performing measurements along discrete survey lines which run vertically in the upwards direction and approximately transversal to the known running direction of the post-tensioning ducts. The measurement equipment was mounted on a frame allowing for automated data collection and a PC with dedicated software was used for visualisation and storage of the recorded data. The operating frequency of the testing equipment as tested on site covered a range from 10 Hz to circa 25 kHz.

One of the impactechograms obtained along one of the vertical survey lines is shown in Figure 17. The horizontal white band (white and paler tones of grey represent greater signal amplitude) at about 7 kHz frequency is one of the main features in the image. It represents the wave reflection at the backside of the web and allows calculating the thickness of the concrete web. Thanks to an estimated wave velocity in the material of 3800 m/s, the concrete thickness was evaluated at 270 mm. From the regular ongoing of this horizontal line it can be deduced that the wave velocity in the concrete is constant, thus that the concrete is homogeneously good. Between stations 10 and 35, a small shift of the thickness frequency towards lower frequencies is noticeable, indicating the presence of a void along the investigated material thickness.

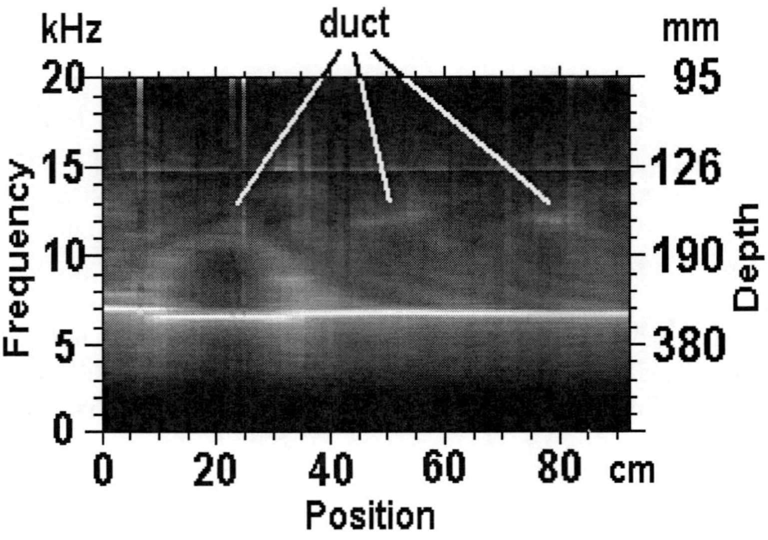

Figure 17. Impactechogram obtained along a measurement line and showing position of ducts.

In addition, two horizontal short segments appear in the frequency range between 11 kHz and 13 kHz. Their lateral position is between stations 45 and 55, and between stations 75 and 85. Their signal intensity is weaker than the thickness frequency, but clearly recognisable in the impactechogram. These two features are interpreted as the presence of two ducts in the concrete [53]. By using equation (13) in section 3.2, their apparent concrete cover was estimated to be 80 mm and 79 mm for the central and right hand side duct respectively. At the position of these two ducts, no shift is registered at the thickness frequency [54-56] thus leading to interpret their grouting condition as good.

On the left hand side of the impact-echogram, it appears, at about 11 kHz, an arch feature with a curved shape, which can be described as a hyperbola turned downwards. Its lateral position corresponds to that where a small shift in the thickness frequency was noticed. This peculiar reflection, previously unrecorded in published literature of impact-echo data from post-tensioning ducts or other structures, constitutes a well recognisable signal patter in the impactechogram. Due to its appearance in combination with a small shift of the thickness frequency towards lower values, this feature is interpretable as a duct with grouting faults [57]. It has an approximate centre at station 22 whilst the hyperbola branches extend widely, from approximately position 0 to 40. By the interpretation criteria formulated by Sansalone and Jaeger [54] and by using equation (2), the depth of this duct was estimated smaller than 82 mm. The horizontal black thin line appearing in the impactechogram at approximately 15 kHz frequency is due to a noise produced by the testing equipment and should be visually ignored.

By assembling the single impactechograms in a 3-D data set it becomes possible to plot a plane area of the tested volume in any desired direction. In particular it is often useful to plot slices parallel to the surface where measurements were carried out, or so called “depth slices”. In these images, the horizontal and vertical axis represent the coordinate positions of the testing unit on the concrete surface, whilst the grey tones represent the amplitude of the reflected wave (Figure 18).

A depth slice from the tested area, at a frequency of 12.3 kHz is shown in Figure 18. It represents a 1 m x 1m section of the concrete web in a plane parallel to the tested surface. This image shows in whitish colour the signal reflections at the depth of the metal tendons in the ducts. The ducts present a running direction sloped on the horizontal line. The bottom duct is less recognizable as the wave reflection occurs at a slightly different depth than the position represented in this plot.

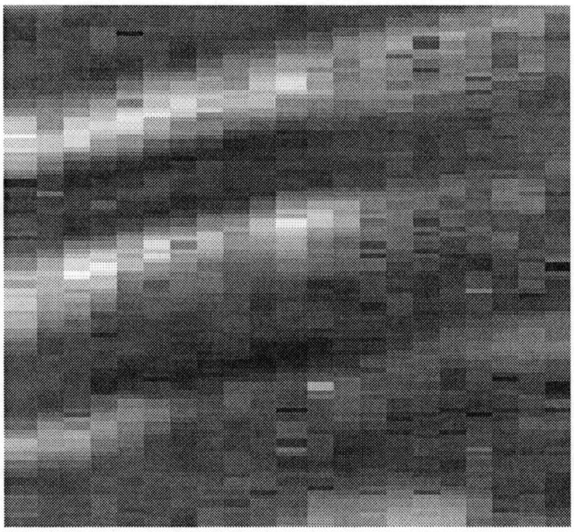

Figure 18. Depth slice of the area investigated via impact-echo and showing in white the positions of the ducts (about 12.3 kHz frequency).

The inspection of the large cores sampled on site showed that the concrete thickness measured between 275 mm and 280 mm: a satisfactory value if compared with the 270 mm estimated via impact-echo. The lateral position of the three ducts as indicated by the analysis of the impact-echo data was found to be correct, with a small error ranging between 5 mm and 10 mm. This value was satisfactory. Observation of the concrete core from the area on the left hand side of Figure 18 showed that the faulty grouting condition of this duct, with a large void between duct and tendons, thus confirming impact-echo data interpretation.

The condition of the middle duct in Figure 18 was verified thanks to a second core: the duct appeared fully grouted, thus again confirming the testing interpretation. Instead, the estimated vertical position of the fully grouted ducts was too conservative with errors ranging around 15-16 mm (20%) and up to 25-30 mm (30% error). This fact shows that the employed equation (2), a modified version of equation (1), failed in providing a satisfactory value of concrete cover for fully grouted ducts and would lead to conservative values of concrete cover. By proposing an alternative equation, concrete cover depths of ducts were re-calculated for the fully grouted ducts and obtained error values were contained between 0 mm and 5 mm thus greatly improving their depth accuracy [58].

In conclusion, the proposed 2-D data imaging facilitates the interpretation of the impact-echo signals collected and allows a more complete structural evaluation. Therefore, the scanning impact-echo is proposed as an advanced and reliable assessment tool for the evaluation of post-tensioned bridges. The quantitative results obtained with this method imply that it is also possible to catalogue the structural damage in accordance to the criteria of the guidelines for construction inspection.

Furthermore, the results of the testing have been made available in 2- and 3-D data visualisation form, which can be easily understood by the engineer responsible for the structure, and related to structural cross-sections.

4.5. Monitoring of Polymer Composite/Concrete Interface Behaviour

An ultrasonic system to monitor the mechanical properties of FRP on concrete is presented in **Error! Reference source not found.** Figure 19.

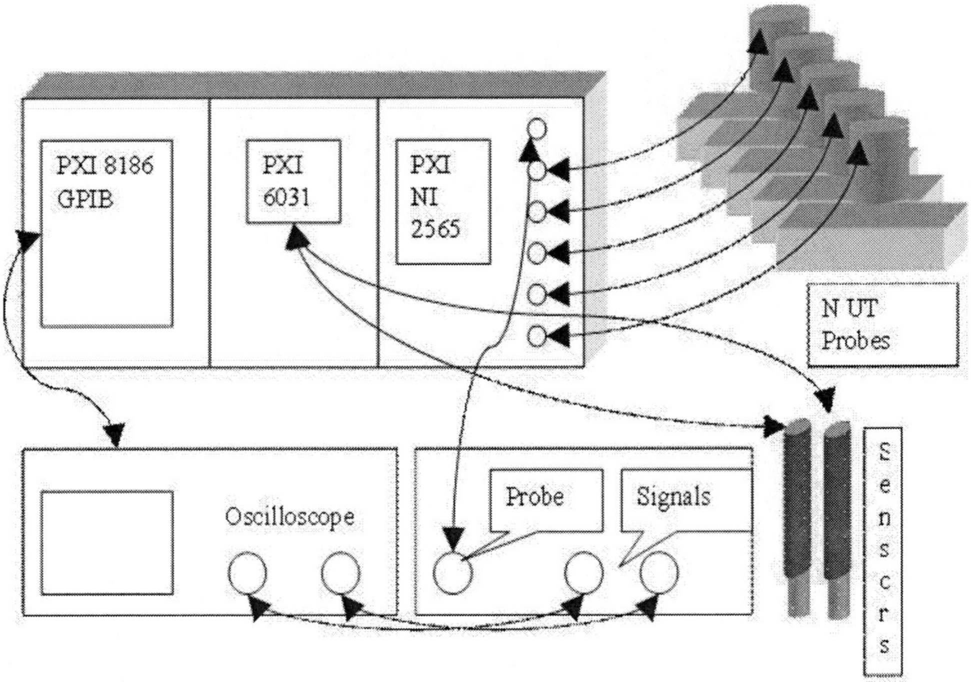

Figure 19. Monitoring system.

Previous studies [60] have demonstrated that it is possible to monitor the progress in the cure reaction of epoxy resins by means of ultrasonic wave and, in turn, the development of mechanical and adhesive strength of bonded structure.

The durability and the performance of FRP-strengthened structures subjected to freezing - thawing cycles (-10° ÷ 40° C) of the FRP itself, as well as that of the FRP-to-concrete interface has been studied during the time.

In particular for four months two samples of concrete with FRP applied with two different techniques as hand lay up and hand lay up plus vacuum bag were monitored with ultrasonic wave. Furthermore, samples of FRP only, epoxy resin (the same used for the composite matrix) and primer were monitored by ultrasound probes (Figure 20), thermocouples and LVDT to measure the thickness variation. In order to obtain information on the chemical - physical behaviour of single constituent of the composite were carried out periodically thermal, mechanical and weight analysis during the four months. At the end of the four months a correlation between mechanical properties measured by means of ultrasonic waves and the other characterizations was established.

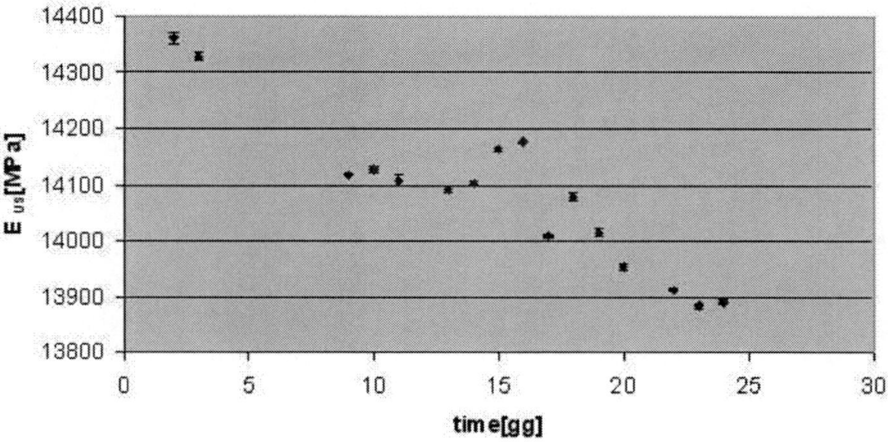

Figure 20. Dynamic Young Modulus of CFRP measured by means of ultrasonic waves monitored during the last month of freezer-thawing cycles.

A good correlation between dynamic modules measured by means of ultrasonic wave and static modulus measured by means of mechanical tensile test has been found. This demonstrates that the monitoring system by means of ultrasound can be a useful instrument to use in situ to monitor variation of mechanical property of CFRP applied to civil structures in concrete.

5. Conclusion

Traditional and innovative non-destructive sonic and ultrasonic techniques in the assessment of existing civil and historic structures have been illustrated in this chapter.

In particular the physical principals and the application of sonic and ultrasonic testing techniques to measure in situ the development in time of mechanical properties of concrete and polymer composite were studied in deep. The assessment of damage and defects in concrete or masonry, and between concrete/masonry and applied polymer composite by means of sonic and ultrasonic techniques through also case studies were illustrated. Different data acquisition procedures and methodologies were described including tomography, scanning acoustic microscope, transmission and surface measurements. Further, an impact–echo example application on post-tensioned concrete was shown. Data interpretation in form of visual reading of signal characteristics from two dimensional images of the tested element has been proposed. The aim was to prove the reliability of image diagnostic in structural applications and in particular for what concerns confidence increase in sonic tests as diagnostic tool for increasing knowledge of the structure under assessment.

References

[1] Ferraro et al. *Res. in Nondestr. Eval.* 2007, *18*, 179-192.
[2] Krishnaiah Chevva et al. *Bull Eng Geol Environ* 2008 *67*, 65–70.
[3] Bond et al. *Construction and Building Materials* 2000 *14*, 133-146.

[4] Schickert, M. *Materials and Structures* 2005, *38*, 807-815.
[5] Buyukozturk , *NDTandE International* 1998, *31*, 233-243.
[6] Vasconcelos, G. et al. *Ultrasonics* 2008, *48*, 453–466.
[7] Kahraman, S. *Ultrasonics* 2007 *46*, 341–348.
[8] UNI-EN 13791 (2005) *Assessment of in-situ compressive strength in structures and precast concrete components*
[9] ACI 228 1R -03, '*In place methods to estimate concrete strength*'
[10] Komlos, K.; Popovics, S.; et al. *Cement and Concrete Composites* 1996, *18*, 357-364.
[11] Krautkramer, *Ultrasonic testing of materials*; Spring-Verlag; Berlin Germany 1977
[12] Perepechko, *Acoustic methods of investigating polymers*, Mir Publishers Moscow (1975)
[13] Ferry, J.D. *Viscoelastic Properties of Polymers*, Lhon Wiley and Sons, New York (1980)
[14] UNI 9524:1989/A1:1992 – 'Hardened Concrete: micro-sismic testing by means of pulsed mechanical waves at high frequency, in concrete sample or structures' - "*Calcestruzzo indurito. Rilievi microsismici mediante impulsi d'onde vibrazionali ad alta frequenza, in campioni o strutture di calcestruzzo semplice, armato o precompresso*"
[15] UNI EN 14579:2005 *Metodi di prova per pietre naturali* - Determinazione della velocità di propagazione del suono
[16] Jackson, M.J.; Tweeton, D.R. *Geophysical tomography using wavefront migration and fuzzy constraints*, R.I. 9497. Bureau of Mines, United States Department of Interior, 1994. 35 p.
[17] Colla, C.; McCann, D.M.; Forde, M.C.; Das, P.C.; Batchelor, A.J. in Proc. VII Int. Conf. *Structural Faults and Repair* -'97, Edinburgh, UK, 8-10 July, 1997, Engineering Technic Press, vol. 1. 143–153.
[18] Binda, L.; Saisi, A.; Tiraboschi, C.; Valle, S.; Colla, C.; Forde, M.J. *Construction and Building Materials*, 2002, 91–106.
[19] Colla, C. in Proc. Italian. Conf. *AIPnD* 2009. Rome, 13-17 October 2009 IDN 85 .
[20] Colla, C.; Benedetti, A.; Pascale, G. in Proc. XXV Conv. Int. *Scienza e Beni Culturali, Conservare e Restaurare il Legno*, Conoscenze, esperienze, prospettive. Bressanone, 23-26 giugno, 2009, Marghera,VENEZIA: Edizioni Arcadia Ricerche. pp. 677–688.
[21] Büyüköztürk, O., *NDTandE International*, 1998, *31*, 233-243.
[22] Ferrero, C.C.; et al. *Res. in Nondestr. Eval.* 2007, *18*, 179-192.
[23] Martin, J. ; et al., *NDTandE International* 2001 *34*, 107-113.
[24] Colla, C. *Non-Destructive Testing of masonry arch bridges.* Ph.D. Dissertation, (1997) The University of Edinburgh, UK.
[25] Carino, N.J.; Sansalone, M.; Hsu, N.N., *Int. Advances in Nondestr. Testing* 1986, *12*, 117-146.
[26] Colla, C.; Krause, M.; Maierhofer, C.; Hoehberger, H.J.; Sommer H. *J. Nondestr. Testing and Eval. Int.* 2002, *35*, 95-105.
[27] Colla, C.; Schneider, G.; Wiggenhauser, H. in: Proc. VIII Int. Conf. *Structural Faults and Repair -'99, CD-Rom Engineering Technics* Press, London, UK, 13-15 July 1999.
[28] Colla, C.; Schneider, G.; Wiggenhauser, H. in Proc. Seiken Symposium No. 26, *NDT in Civil Engineering*, Uomoto, T. (ed.);Tokyo, Japan, 25-27 April 2000, Elsevier 2000, pp. 471-480.

[29] Ghorbanpoor, A. *Evaluation of post-tensioned bridge structures by the impact echo technique.* Publication No. FHWA-RD-92-096. Federal Highway Administration. U.S.A. 1993.

[30] Krieger, J.; Krause, M.; Wiggenhauser, H. in Proc. SPIE *Structural Materials Technology III* Medlock, R.D. and Laffrey, D.C. (Eds.), 1998, vol 3400 pp. 258-269.

[31] Martin, J.; Hardy, M.; Usmany, A.; Forde, M.C. in: Proc. 7th Int. Conf. *Structural Faults and Repair*, Edinburgh, UK, July 1997, vol. 1, 341-354.

[32] Sack, D.A.; Olson, L.D. *NDTandE International* 1995, *28*, 349-357.

[33] Sansalone, M.; Carino, N. *J. ACI Materials* 1989, 175-184.

[34] Sansalone, M.; and Carino, N. "Impact-echo: a method for flaw detection in concrete using transient stress waves", National Bureau of Standards Report NBSIR 86-3452, Gaithersburgh, Maryland, September 1986.

[35] Sansalone, M.; Streett, W. in Proc. Int. Symp. *Non-destructive Testing in Civil Engineering (NDT-CE)*, Berlin 1995, pp. 495-502.

[36] Sansalone, M. J. and W. B. Streett: Nondestructive evaluation of concrete and masonry. Bullbrier Press. Ithaka. N.Y. 1997.

[37] Mal, A.; Ricci, F.; Banerjee, S.; Shih, F. *Structural Health Monitoring-an International Journal* SEP 2005, *4*, 283-293.

[38] Kolluru, V.; Subramaniam, Jaejun Lee - International RILEM Symposium on *Concrete Science and Engineering* A Tribute to Arnon Bentur Editor(s): J. Weiss et al. 2004.

[39] Robeyst, N. et al., *Cement and Concrete Research* 2008, *38*, 1169–1176.

[40] Lei Qin; Zongjin Li , *Smart Mater. Struct.* 2008, *17*, (6pp)

[41] Franklin R. et al. *Nondestr.Test.Eval.* 2001, *17*, 59-78.

[42] Luprano, V. et al., in Proc. Int. Symposium *Polymers in Concrete* 2006 vol.1, pp. 341-353.

[43] Nanni, A. *Fiber-reinforced-plastic (FRP) reinforcement for concrete structures: properties and applications*, Elsevier, Amsterdam, (1993)

[44] Van Den Einde L, Zhao L, Seible F., *Construction and building materials* SEP-OCT (2003) 17 (6-7), 389-403

[45] Karbhari, V.M.; Chin, J.W.; Reynaud, D. *Journal of Composites for Construction*, ASCE 2003, *7*, 241-242.

[46] Luprano V.; Pfister, V.; Tundo A. *Construction and Building Materials* (submitted)

[47] OPCM 3274, 2003, '*Primi elementi in materia di criteri generali per la classificazione sismica del territorio nazionale e di normative tecniche per le costruzioni in zona sismica*', Gazzetta Ufficiale 8/5/2003

[48] Colla, C.; de Miranda, S.; Ubertini, F. in Proc. RILEM Int. Conf. SACoMaTiS *On site Assessment of Concrete, Masonry and Timber Structures*, September 2008 Varenna (LC), Italy.

[49] Colla, C.; Pascale, G. Italian Conference on *Non Destructive Testing, Monitoring and Diagnosis*, Milan October 2007 (in Italian language).

[50] BA 16/93. (1993) *The assessment of highway bridges and structures*. HMSO, London 1993.

[51] Das, P.C. (1993). Symp on *Structural Preservation of the Architectural Heritage*. IABSE, Rome, 1993.

[52] Colla, C.; Das, P.C.; McCann, D.; Forde, M.C. in: Proc. III Int. Symp. *Non-Destructive testing in Civil Engineering (NDT-CE)*, September 1995, Berlin, Germany, DGZfP, vol. 1, p. 629-636
[53] Colla C. "*Impact-echo", Materialtechnische Untersuchungen beim Abbruch der Talbruecke Haiger*, M. Krause, H. Wiggenhauser and J. Krieger (Ed.), 29-57, BAM-BASt internal report, Berlin- Bergish Gladback, v. July 2001
[54] Jaeger B.; Sansalone, M.; Poston, R. *Structural Journal* 1996, *93*, 462-474.
[55] Jaeger B.; Sansalone, M.; Poston, R. *Concrete International* 1997, 42-46.
[56] Sansalone,M.; Streett, W. Impact-Echo, nondestructive evaluation of concrete and masonry, 340 pp., Bullbrier Press, Ithaca, N.Y., 1997.
[57] Colla, C. in Proc. 1st Int. Conf. on *Bridge Maintenance, Safety and Management, IABMAS* 2002, Edited by J.R.Casas, D.M. Frangopol and A.S. Novak, Barcelona,CIMNE, Barcelona, book of abstracts and CD-Rom.
[58] Colla, C., in Proc. 29th annual *Review of Progress in QNDE* Conference, Bellingham, WA, July 14 - 19, 2002. Edited by D. O. Thompson, D.E. Chimenti, American Institute of Physics, 657,Melville, New York, 2003, Vol. 22B, pp. 1185-1192.
[59] Colla, C. in Proc. of *SMT Structural Materials Technology: NDE/NDT for Highways and Bridge Topical*, S. Alampalli and G. Washer (Ed.), Cincinnati, OH, September 2002, ASNT, Columbus, pp. 307-312.
[60] Maffezzoli, A. et al. *Journal Applied Polymer Science* 1999, *73*, 1969-1977.
[61] Frigione M. et al. *Polymer Engineering and Science* 2000, *40*, 656-664.
[62] Frigione, M.; Maffezzoli, A.; Luprano, V. *Journal of Polymer Engineering* 2001, *21*, 445-468.

In: Recent Advances in Non-Destructive Inspection
Editor: Carosena Meola, pp. 89-123
ISBN 978-1-61668-550-8

Chapter 4

INFRARED THERMOGRAPHY IN NON-DESTRUCTIVE INSPECTION: THEORY AND PRACTICE

Carosena Meola and Giovanni Maria Carlomagno
Department of Aerospace Engineering (DIAS)
University of Naples Federico II, Via Claudio, 21, 80125 Napoli, ITALY

ABSTRACT

The attention of this chapter is focused on the use of infrared thermography for non-destructive evaluation of materials. First, the historical development of the technique is traced going back to the discovery of thermal energy; some basic theoretical concepts are also recalled. Then, the use of two main non destructive techniques which are pulse thermography (PT) and lock-in thermography (LT) is illustrated through theory and practical examples. A list of applications with reference to literature is supplied where infrared thermography is usefully exploited. Such applications include different types of materials (glass and carbon reinforced polymers, sandwiches and fibre metal laminates), masonry structures, and metals involving different types of anomalies such as delamination, slag inclusions, cracks, impact damage. At last, some results obtained from inspections of materials, which were performed at the University of Naples are presented and discussed.

INTRODUCTION

Infrared Thermography (IRT) is a methodology which allows for: remote detection of thermal energy that is radiated from objects in the infrared band, conversion of such energy into a video signal, and representation of the object surface temperature map. This represents a great potentiality to be exploited in a lot of application fields and for many different purposes. For example, IRT may be used for diagnosis (in medicine, architecture, maintenance), or for understanding of complex fluid dynamics phenomena (flow instability, flow separation and reattachment), or for material characterization and procedures assessment which can help improving design and fabrication of products. As technology evolves, infrared systems offer new opportunities for application; indeed, any process which is temperature-dependent may benefit from the use of an infrared device.

Infrared thermography may accompany the entire life of a product, from manufacture (on-line process control) [1-2], to final product (nondestructive evaluation) [1-17] and to in-service (maintenance) [3-5]. What is important, is the choice of the most adequate system for a specific application. Trends in the IRT history and application in three fields such as fluid-dynamics, technology, and cultural heritage may be found in review works mainly those by Meola and Carlomagno [1, 17], Vavilov [7], and Maldague [16] and in the Proceedings of the main Conferences concerned with IRT and which are the Quantitative InfraRed Thermography QIRT (biennial starting from 1992), the Advanced Infrared Technology and Applications AITA (biennial starting from 1991), SPIE Thermosense Conferences (annual starting from 1978), the Inframation (from 1999).

The aim of this book is an overview on non destructive inspection, and then this chapter is limited to the use of infrared thermography as non destructive technique. Signs of the use of infrared thermography in non-destructive inspection were traced by Vavilov [7] up to the beginning of the last century, when the researchers efforts were mainly driven by military purposes. It was in the 60-s that the IR military applications started to penetrate the civil engineering areas [18-21]. The take and try approach went ahead for some years and to the end of the 70-s it seemed exhausted without successful competition with other methods.

A new stimulus came from heat transfer basics which provided understanding of the technique mechanism and so starting from the 80-s till now there has been a continuous proliferation of ideas and/or procedures to improve the technique. An up-to-date panorama can be found in the proceedings of two main specific international conferences, namely the European QIRT (Quantitative Infrared Thermography) and the American Thermosense. Information about the use of infrared thermography as non-destructive technique can be also found in Maldague [16] (2001), Meola and Carlomagno [17] (2006), Dillenz et al. [22] (2003).

1. A Quick Look on the Bases of Infrared Thermography

The bases of infrared thermography dates back to the early 1800s when the German-English astronomer Friedrich Wilhelm Herschel (better known as Sir William Herschel) discovered thermal radiation beyond the deep red in the visible spectrum, the *invisible light* later called *infrared*. From this observation and thanks to the studies of many other scientists (amongst them Macedonio Melloni, Gustav Kirchhoff, Clark Maxwell, Joseph Stefan, Ludwig Boltzmann, Max Planck, Heinrich Rudolf Hertz, Theodore Willard Case), in the successive years, it was possible to develop infrared imaging devices.

An infrared imaging system basically includes a camera, and a computer, which, depending on the model, may be incorporated inside the camera itself. Of course, the core of the camera is the infrared detector, which senses the IR energy emitted by the object (whose surface temperature is to be measured). Indeed, every object, which is at a temperature above absolute zero, emits thermal energy; this energy is captured by the infrared detector, transformed into an electronic video signal and ultimately into a temperature map. In fact, each energy level (temperature) can be represented by a colour, or a grey level and such a map is generally called thermogram, or thermal image [7-8]. However, the amount of emitted energy not only depends on the object temperature, but also on the emissivity of the surface under measurement and on the environmental temperature.

The law which describes the spectral distribution of the radiation intensity from a blackbody was derived by Planck:

$$E_{\lambda b} = \frac{C_1}{\lambda^5 (e^{C_2/\lambda T} - 1)} \quad (1)$$

where: $E_{\lambda b}$ is the blackbody monochromatic radiation intensity, C_1 and C_2 are the first and second radiation constants respectively, λ is the wavelength of the radiation being considered and T is the absolute temperature of the blackbody. The electromagnetic spectrum is roughly divided into a number of wavelength regions called *bands.* The infrared spectral band is generally sub-divided into lesser bands, also called infrared windows, (with arbitrarily chosen boundaries): *Near Infrared* (NIR), *Middle Wave Infrared* (MWIR), *far (or Long Wave) Infrared* (LWIR) and *extreme infrared.* A sketch of the electromagnetic spectrum is given in Figure 1.

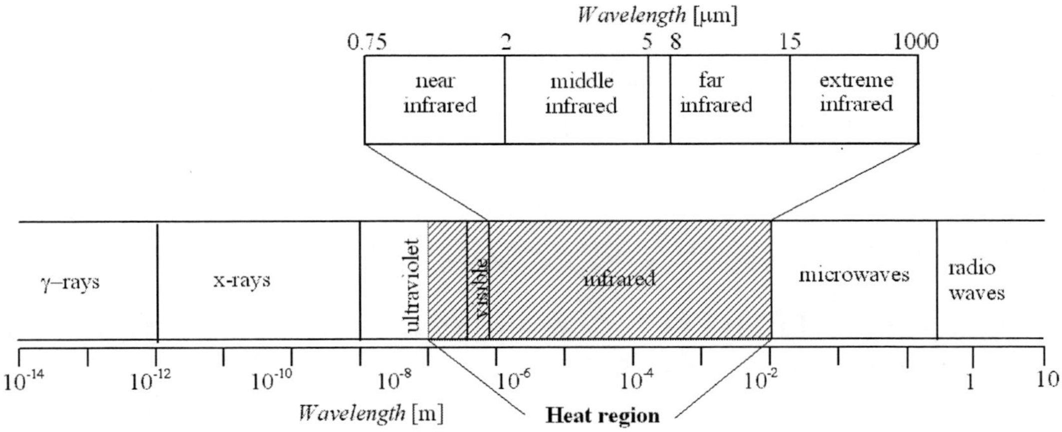

Figure 1. Representation of the electromagnetic spectrum.

By differentiating the Planck's law with respect to λ and looking for the maximum radiation intensity the Wien's displacement law is obtained:

$$\lambda_{max} = 2898/T \quad (2)$$

which mathematically expresses the common observation that colours vary from red to orange to yellow, as the temperature of a thermal radiator increases. So, the sun (about 6000 K) emits yellow light peaking at about 0.5 μm in the visible light spectrum. At ambient temperature (300 K) the peak of radiation lies in the far infrared at about 10 μm, while at the temperature of liquid nitrogen (77 K) the maximum of the almost insignificant radiation intensity occurs at 38 μm in the extreme-infrared wavelength. The Planck's formula is plotted in Figure 2 for several absolute temperature values in the range 200-6000 K; the Wien's displacement law is also represented.

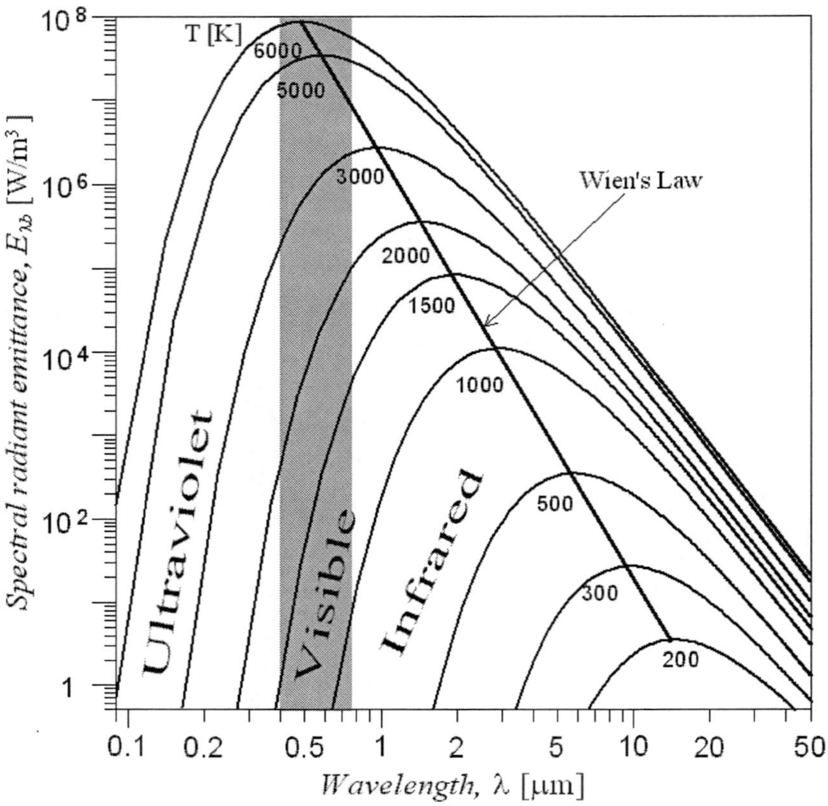

Figure 2. Monochromatic emissive power versus wavelength for varying temperature.

By integrating the Planck's law over the entire spectrum (λ = 0-∞), the total hemispherical radiation intensity is obtained:

$$E_b = \sigma T^4 \tag{3}$$

where σ is the Stefan-Boltzmann constant.

It has to be pointed out that Eq.(1) describes the radiation emitted from a blackbody which is the maximum value radiated by a body at a given temperature. Real objects almost never comply with this law although they may approach the behaviour of a blackbody in certain spectral intervals. A real object generally emits only a part E_λ of the radiation emitted by a blackbody at the same temperature and at the same wavelength. By introducing the quantity:

$$\varepsilon_\lambda = \frac{E_\lambda}{E_{\lambda b}} \tag{4}$$

which is called spectral emissivity coefficient, Eq.(1) can be rewritten for real bodies by simply multiplying its second term by ε_λ. The energy really sensed by the detector depends not only on the emissivity coefficient of the surface under measurement but also on the environment. In fact, as represented in Figure 3, a fraction may be either absorbed, or emitted, by the atmosphere between the object and the camera, either added as reflected by the surface from the surroundings. Infrared detectors sense radiation in a limited bandwidth of the IR spectrum. Measurements are generally performed in two windows: MWIR (2-5.5 μm) and LWIR (8-12 μm); systems which work in the NIR window are tailored for specific applications. Between 5.5 and 8 μm no measurements are performed since the atmosphere, in that region, practically acts as opaque substance; no (or very little) infrared radiation is transmitted throughout.

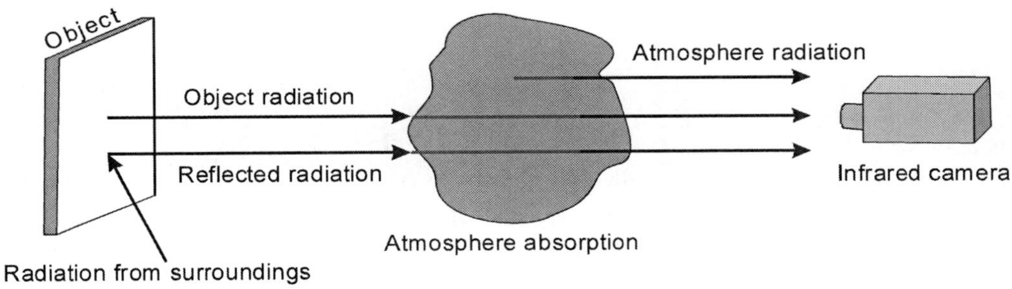

Figure 3. Sketch of the atmosphere influence on the radiation captured by the infrared camera.

Infrared thermography can be successfully exploited in many industrial and/or research fields; obviously, each field presents specific characteristics and requirements which entail specific choice of the infrared system, the test procedure and the successive data analysis.

2. Infrared Systems

Infrared instruments are often wrongly classed as total-radiation radiometers and are considered to be based on the Stefan-Boltzmann's law even if their detectors sense radiation in only a limited bandwidth of the IR region. Actually, infrared devices perform measurements in three main IR bands: *Near Infrared* NIR (0.9-1.7μm), *Middle Wave Infrared* MWIR (2-5.5 μm) and *Long Wave Infrared* LWIR (8-14 μm). The atmosphere is practically opaque to infrared in the region between 5.5 and 8 μm and thus no measurements are possible in that region. Therefore, measurements made with infrared radiometers are more correctly based on the Planck's law.

The ultimate result of an infrared system is a surface temperature map. However, up to a few years ago, the sensing element was zero-dimensional (practically a point). Two-dimensionality was achieved by a scanning mechanism which consisted of oscillating mirrors, or rotating refractive elements (such as prisms), and allowed object scanning in both vertical and horizontal directions. Nowadays, the new systems are based on the Focal Plane Array (FPA) technology; but, often they are still referred to as Infrared Scanning Radiometers (IRSRs) since their output comes from an electronic scanning.

The overall performance of an IR imaging system is conventionally evaluated in terms of useful and accurate information that can be acquired per unit time. This can be expressed through several parameters such as thermal sensitivity, or equivalent random noise level; scan speed, or update rate of the scanning mechanism; image resolution, or number of independent measurement data points of which the image is composed; and intensity resolution, or number of intensity levels, which allows for fine temperature differences, i.e. dynamic range.

The core of the IR thermographic system is the detector (or array of detectors); basically, it is a transducer that absorbs the IR energy emitted by the object (being measured) and converts it into a signal, usually an electrical voltage or a current. In the following, some pieces of information about infrared detectors are supplied for the chapter completeness; for more insights the reader is addressed to specific literature [23-43].

2.1. Detectors Classification

Detectors can be grouped into two main categories: *thermal* and *quantum* (or photon) detectors and can be classified as *cooled* and *uncooled* infrared detectors.

Cooled detectors require cooling well below the ambient temperature to allow for rapid scanning, high sensitivity and low noise. In the past, the detector was often located in the wall of a dewar flask which was filled with liquid nitrogen, LN_2 (77 K). The new generation systems include miniature coolers based on the thermoelectric Peltier effect or the Stirling cycle and do not require any external cooling source.

Uncooled infrared detectors operate at ambient temperature, or at a temperature close to ambient using small temperature control elements. These detectors work generally by the change of a material property (resistance, voltage or current) when heated by infrared radiation. These changes are then measured and compared to the values at the operating temperature of the sensor.

2.1.1. Thermal Detectors

These detectors are also called energy detectors, or photon absorber because they absorb the incident energy and warm up; the temperature changes are measured through the variation of a temperature-dependent property of the material such as the electrical resistance. They perform well at room temperature and over a large range of the electromagnetic spectrum. The main disadvantage is a relatively slow response time, which depends on the heat capacity of the detector material and on the wavelength of the incident radiation. Examples of detectors that belong to this family are: thermopiles, bolometers, pyroelectric detectors, microcantilevers.

Thermopiles (based on the thermoelectric effect). A thermopile is obtained by a number of thermocouples connected in series. As well known a thermocouple is obtained by joining two dissimilar materials. Heating the junction causes a voltage (Seebeck effect) that is proportional to the temperature change and thus the temperature may be computed through the measured voltage. The earliest thermopiles were built in a simple way by using fine wires thermocouples (e.g., copper-constantan). The temperature of the hot junctions increased until an equilibrium was reached between the rate of incident energy and the loss by conduction, convection and radiation to the surroundings. The thermoelectric output was measured by a suitable device and transformed into temperature (i.e., the temperature of the target). In the modern technology thin films are deposited on a substrate; this allows a reduction of dimensions, the sensitivity can also be improved by encapsulating the thermopile in a low thermal conductivity medium (i.e., avoid losses to the surroundings).

Bolometers (change of resistance with temperature). The bolometer is a resistor element; the basic design consists of two platinum strips, covered with lampblack, one strip is shielded from the radiation and one exposed to it. The strips form two branches of a wheatstone bridge; the resistance in the circuit varies when electromagnetic radiation, falling on the exposed strip, heats it up and changes its electrical resistance. Nowadays, advances in micromachining of silicon have lead to the microbolometer technology which includes a grid of vanadium oxide or amorphous silicon heat sensors atop a porous silicon bridge as thermal isolating and mechanical supporting structure. The microbolometer grid is commonly found in two sizes, a 320×240 array or less expensive 160×120 array.

Pyroelectric detectors (change of surface charge with temperature). The pyroelectric effect consists of a change in the crystal surface charge as a consequence of the change in the dipole moment when the crystal temperature is raised. For continuous operation, the common technique is to incorporate a chopping device inside the optical system and to create an AC output signal; the spurious signals caused by low-frequency ambient temperature can be filtered. Triglycine sulfate (TGS), lithium tantalate (LiTaO3) and polyvinyl fluoride (PVF2) are materials that exhibit the pyroelectric effects.

Microcantilevers (bimetal bending effect). These detectors are based on the dissimilar thermal expansion of bimetals (e.g., silicon nitride and gold film). More specifically, a cantilever forms a capacitance in combination with a reference plate; a temperature increase, due to exposition to infrared radiation, causes bending of the cantilever which, in turn, alters the capacitance of the structure. The intensity of IR radiation is measured as being proportional to the bending extension.

2.1.2. Photon detectors

Photon detectors generate free electrical carriers in response to photon absorption. The main advantage is a very short response time (of the order of microseconds). A disadvantage is the need of cooling down to cryogenic temperature to get rid of excessive dark current. In fact, electrons can be excited from the valence to the conduction band by photons having an energy E_{ph} that is larger than the energy E_g between the bands; there exists a wavelength λ_c named cut-off wavelength which corresponds to the energy E_g:

$$\lambda_c = \frac{hc}{E_g} \tag{5}$$

beyond which no emission occurs. The value of E_g varies from semiconductor to semiconductor; it tends to increase as temperature increases (wavelength decrease). Therefore, the detector must be maintained at low temperature.

There are four main types of photon detectors: photovoltaic, photoconductive, photo-emissive and Quantum Well Infrared Photodetector (QWIP). In particular, the photovoltaic and photoconductive types may be intrinsic, or extrinsic.

Photovoltaic intrinsic (electromotive force generation). The structure of a photovoltaic (PV) detector is based on a p-n junction device (two dissimilar materials) which, under IR radiation, generates photocurrents (current flows across the junction of the two materials). More specifically, when the detector is knocked by photons of energy greater than or equal to the energy band-gap, electrical carriers are swept across the photodiode and become separated by the potential barrier of the p-n junctions to create a current. PV devices operate in the diode's reverse bias region; this minimizes the current flow through the device which in turn minimizes power dissipation. These detectors are generally fabricated from silicium (Si), germanium (Ge), gallium arsenide (GaAs), indium antimonide (InSb), indium gallium arsenide (InGaAs), and mercury cadmium telluride (HgCdTe) also called (MCT).

Photoconductive intrinsic (conductance change). In photoconductive (PC) detectors, the incident radiation, with energy greater than or equal to the energy band-gap of the semiconductor, generates majority electrical carriers. The electric conductivity of the material is improved; there is an internal photoelectric effect. Common materials are lead sulfide (PbS), lead selenide (PbSe) and mercury cadmium telluride (HgCdTe).

Extrinsic detectors. Extrinsic detectors are similar to intrinsic detectors. The difference lies in the fact that carriers are excited from the impurity levels and not over the band-gap of the basic material; this is achieved by doping the semiconductor material.The mostly used materials are silicium and germanium doped with impurities such as boron, arsenic and gallium. The spectral response of these detectors can be controlled by the doping level.

Photo-emissive detectors. These are bi-materials detectors, a layer of a metal over a layer of a semiconductor; a typical example is platinum silicide (PtSi) on silicon (Si). The process, which is also called external photoelectric effect, consists in the emission of carriers from the metal into the semiconductor under photon absorption. The main advantage is a more uniform response since the response depends on the characteristics of the metal. However, being the photon absorption proportional to the square of the wavelength, this type of detector is more indicated for long wavelengths.

Quantum Well Infrared Photodetector (QWIP). The basic principle is similar to that of extrinsic detectors with the peculiarity that the dopants are concentrated in microscopic regions and creates the quantum wells. The radiation is absorbed by the entire quantum well, not only by a single dopant atom, and thus the absorption is increased (and also the response) with respect to extrinsic detectors. QWIPs generally consists of layers of thin gallium arsenide (GaAs) alternated to layers of aluminum gallium arsenide (AlGaAs).

2.2. Detectors Performance

The performance of a detector is evaluated through three parameters that are: Responsivity R_v, Noise Equivalent Power (NEP) and Detectivity D^*.

Responsivity The responsivity R_v is a measure of the signal output S_{out} (voltage, or current) per incident radiation E_{inc} over the active area of the detector A_{det}:

$$R_v = \frac{S_{out}}{E_{inc} A_{det}} \tag{6}$$

The active area is determined as the ratio $A_{det} = \frac{A_o}{A_e}$ between the optical area A_o and the actual *electrical* area A_e. The responsivity can also be written as:

$$R_v = \frac{\lambda \eta_q Q_e p_g}{hc} \tag{7}$$

with η_q the quantum efficiency, Q_e the electron charge, p_g the photoelectric gain.

Noise Equivalent Power This parameter (NEP) defines the intrinsic noise level of the detector, or better the detection limit of the detector. It can be expressed by:

$$NEP = \frac{E_{inc} A_{det}}{\frac{S_{out}}{N_{out}} \sqrt{\Delta f}} = \frac{N_{out}}{R_v \sqrt{\Delta f}} \tag{8}$$

with N_{out} the noise output and Δf the noise bandwith.

Detectivity The Detectivity *D** defines the resolving power of the detector and is expressed in terms of the signal-to-noise ratio (S_{out}/N_{out}) with respect to the incident power:

$$D^* = \frac{S_{out}}{N_{out}} \frac{1}{E_{inc}} \frac{\sqrt{\Delta f}}{\sqrt{A_{det}}} = \frac{\sqrt{A_{det}}}{NEP} \tag{9}$$

The common units of *D** are cm $Hz^{1/2}/W$; often, in literature, this quantity is measured in Jones in honour to D. Clark Jones who defined this magnitude in 1959. The noise that affects detection of IR radiation comes from two main sources: the IR detector itself with its circuits and the background fluctuations. The first contribute is generally negligible with respect to the second one and then the detection limit is evaluated by accounting only for the background radiation. This contribution is called Background Limited Infrared Photodetection (BLIP). Photovoltaic detectors are characterised by a $\sqrt{2}$ higher BLIP; in fact, the relationship for a photovoltaic detector is :

$$D^*_{BLIP} = \frac{\lambda}{hc} \frac{\sqrt{\eta_q}}{\sqrt{2\Phi_b}} \tag{10}$$

while for a photoconductive is:

$$D^*_{BLIP} = \frac{\lambda}{2hc}\frac{\sqrt{\eta_q}}{\sqrt{\Phi_b}} \tag{11}$$

Φ_b being the background photon flux density (also called dark current). The value of Φ_b received by the detector depends on its responsivity to the wavelengths contained in the radiant source and on its Field Of View (FOV) of the background.

A relevant parameter for infrared systems is the Noise Equivalent Temperature Difference (NETD); it can be defined as the temperature change, for incident radiation, that gives an output signal equal to the r.m.s. noise level N_L:

$$NETD = N_L \frac{\Delta T}{\Delta S_m} \tag{12}$$

ΔS_m is the signal measured for the temperature difference ΔT.

2.3. Progress in IR Detectors Technology

The first infrared detector may be considered the thermometer used by Herschel; i.e., the simple mercury thermometer that responds with a temperature variation to the incident radiation. The history of infrared detectors for remote sensing really originates with the thermoelectric effect at the junction of two dissimilar metals, discovered by the Estonian Thomas Johann Seebeck in 1821. This effect, which is also called Seebeck effect, consists in a direct conversion of temperature differentials into electric voltage and allowed the invention of the thermopile by Nobili in 1829. Later, in 1880, the property of metals to change electrical resistivity with temperature variation lead to the invention of the bolometer by the American Samuel Pierpont Langley.

A milestone in detectors development is certainly the photoelectric effect, which was firstly observed by Hertz in 1887, and explained and formulated by Einstein in the early 1900s. Closely related to the photoelectric effect is the photoconductive effect that is the increase of electrical conductivity of non metals when exposed to electromagnetic radiation.

The first phoconductive-based detector, a thallous sulphide, was developed by Theodore Willard Case in 1917. Such a detector suffered from problems of noise and stability; it was subject of study during World War II and later dropped in favour of lead sulphide (PbS) which had higher performance. Indeed, from the late 1940s and continuing in 1950s, new materials were developed such as lead selenide (PbSe) and lead telluride (PbTe) which allowed a shift of spectral sensitivity from the (1-3μm of PbS) to (3-5μm) infrared wavelength. The end of the 1950s and the beginning of the 1960s saw the introduction of semiconductor alloys, such as the indium antimonide (InSb) and the mercury cadmium telluride (HgCdTe) or (MCT), which allowed the spectral response to be custom tailored from mid (MWIR) to long (LWIR) wavelength infrared. MCT was at once considered the most important and versatile material for IR detector applications; indeed, this material has been and is still now under study. MCT allows for both photoconductive and photovoltaic detection and has inspired the development of the three *generations* of detector devices. Initially, some difficulties arose in producing high-uniformity detectors. One main negative aspect was the weak Hg-Te bond which entailed instability. Therefore, initially, single element detectors were produced and the scenery image was obtained through a mechanical scanning mechanism. These detectors belong to the so-called *pre-first generation*.

The *first generation* involved linear arrays. However, the fabrication of arrays of sensors started with the development of photolithography in the early 1960s; PbS, PbSe and InSb detectors were first used in linear array technology. In the late 1960s and beginning of 1970s linear arrays of intrinsic MCT photoconductive detectors were developed. As technology further advanced, linear arrays of 60, 120, or 240 pixels were produced with acceptable uniformity and called linear focal plane arrays (FPA). The video image was obtained by scanning, through a rotating mirror, the linear arrays along with a line-filling scheme in the vertical axis. The resulting image had a quite good resolution (comparable to TV standards) in the horizontal direction and a low one in the vertical direction. IR systems with detectors of the first generation are still in use.

The *second generation* consists of two dimensional (2D) arrays of detectors that are now in wide production. The basic idea of this type of detector originated from the invention of the charge coupled device (CCD) in 1969 by Willard Boyle and George Smith. It was possible to have detection and readout implemented on one silicon chip. The detectors are configured in a 2D array generally referred to as *staring* FPA. These arrays are electronically scanned by readout integrated circuits (ROICs). The second generation is a full-framing system and has a great number of pixels (about three orders of magnitude) than the first generation. There are also intermediary systems which include linear arrays scanned with multiplexed devices; of course, these systems are characterized by a time delay integration (TDI).

FPAs can be developed in different architecture and generally grouped into: *hybrid* FPAs and *monolithic* FPAs. The monolithics, as the term lets to suppose, are compact with both the IR sensitive material and the signal transmission paths on the same layer. The basic element in a monolithic approach is a metal insulator semiconductor (MIS) with the main function of charge transfer device. In other words, a MIS detects and integrates the IR generated photocurrent (multiplexing function). This results in a more simple and low-cost manufacture, but with the penalty of a low fill factor (i.e., the ratio of active IR sensing material to inactive row and column borders) and consequently low sensitivity with respect to their hybrid counterpart. In hybrid FPAs, multiplexing can be achieved with two different circuits the CCD and the complimentary metal oxide semiconductor (CMOS).

The *third generation* essentially includes large format multicolour and avalanche photodiodes (APD). A significant advantage of multicolour (or multi spectral) technology is their capability of sensing in two or more spectral bands; this can provide added information that is not available with traditional single-colour systems such as, with the assumption of gray emitting body, to avoid to know the emissivity coefficient to determine its temperature. A primary requisite of third generation detectors (TGD) is to provide high spatial resolution (of the order of 1×10^7 pixels against the maximum of 3×10^6 of the second generation staring FPAs), which also means reduction of pixel pitch. Indeed, the goal of TDG is to achieve high frame rates (simultaneous or sequential reading) and better thermal resolution with very low NETD of about 1 mK against the 20-30 mK of FPAs. As demonstrated by Rogalski [24] the

NETD is inversely proportional to the square root of the integrated charge and then a low NETD value requires a high charge storage capacity. This is not possible with standard CMOS capacitors [23].

Competing Technologies

The $Hg_{1-x}Cd_xTe$ ternary alloy [24, 25] is considered as the near ideal material since it can be tuned from SWIR to LWIR by adjusting the x composition of the alloy. The HgCdTe for third generation detectors involves a novel technology based on molecular beam epitaxy (MBE), which may be germanium (mainly used by the French SOFRADIR), or silicon (mainly used in USA by Raytheon Vision Systems (RVS)). The connection with indium bumps to a ROIC chip as a hybrid structure, also called sensor chip assembly (SCA), allows for very large array manufacturing.

Research organizations are working for the development of large multi-band IR detectors. Among them, in USA the RVS has developed a 1280×720 pixel dual-band MW/LWIR FPA under the U.S. Army's Dual-Band FPA Manufacturing (DBFM) program. Such dual-band FPA architecture was fabricated from MBE-grown HgCdTe triple-layer heterojunction (TLHJ) wafers and is sponsored to provide highly simultaneous temporal detection in the MWIR and LWIR bands using time-division multiplexed integration (TDMI) incorporated into the ROIC. In France, the SOFRADIR is involved with the CEA-LETI/LIR within a Design of Excellence for the Future of Infrared (DEFIR). In the LWIR region, the HgCdTe material is limited by metallurgical problems of the epitaxial layers such as uniformity and number of defective elements [26].

A low-cost alternative technology is the QWIP. A 1024×1024 QWIP pixel FPA working in MWIR and LWIR has recently been developed and tested by a NASA lead-team. This is a multi-quantum-well structure which consists of coupled quantum wells of GaAs, InGaAs thin (10, 20Å) doped layers and un-doped AlGaAs barriers [27, 28]. QWIP dual band detectors provide good resolution but involves integration times of the order of 5-10ms while, for fast changing scenes, a lower integration time (typically 1ms) is required.

In this context, antimonide based type II superlattices have emerged as third candidate for third generation infrared detectors [23, 29]. Type II InAs/GaInSb superlattice structure has great potential for LWIR and very LWIR spectral ranges it is a mechanically robust III–V material and has a direct band gap with wavelength tunability from 3 to 25 μm [26]. A major advantage is the higher potential performance compared to HgCdTe with the same cut-off wavelength. The lower production costs are a good premise for commercial market applications in the future. However, they are an early stage of development and problems still exist in material growth, processing, substrate preparation, and device passivation.

3. Non Destructive Inspection of Materials

Non destructive inspection of materials with an infrared imaging system generally consists in the thermal stimulation of the object (under exam) and monitoring of its surface temperature variation during the transient heating, or cooling, phase. The analysis is performed by relating the latter one to the heat conduction in solids [44]. The presence of an inhomogeneity in a material affects the heat propagation and causes a local surface temperature variation.

3.1. Temperature Variation Under Heat Propagation In Solids

The general equation governing the temperature variation under thermal stimulation of a solid is the Cattaneo's equation [45]:

$$\dot{q}+t_r\frac{\partial\dot{q}}{\partial t}=-k\nabla T \tag{13}$$

with $\dot{q}$ the heat flux, T the temperature, t the time and t_r and k the relaxation time and the thermal conductivity. Eq. (13), applying the conservation of energy (first thermodynamics law), can be rearranged to have the hyperbolic (telegraph) equation:

$$\frac{1}{V_h^2}\frac{\partial^2 T}{\partial t^2}+\frac{1}{\alpha}\frac{\partial T}{\partial t}-\nabla^2 T=0 \tag{14}$$

with $\alpha=k/\rho c_p$ the thermal diffusivity coefficient, ρc_p the volumetric specific heat and V_h the speed of propagation of thermal waves.

$$V_h=\sqrt{\frac{\alpha}{t_r}} \tag{15}$$

if $V_h\to\infty$, or $t_r\to 0$, i.e., the case of zero relaxation time and infinite heat propagation velocity, the hyperbolic equation (14) becomes the parabolic Fourier's heat conduction equation:

$$\frac{\partial T}{\partial t}-\alpha\nabla^2 T=0 \tag{16}$$

The heat propagation through a material is driven by its thermal properties. Considering a slab of thickness s as sketched in Figure 4. Such a slab, initially ($t = 0$) is at uniform ambient temperature (T_a), which is to say $T_1=T_2=T_a$, is exposed from side 1 to a uniform heat source for a certain time period. Two cases may arise depending on the slab characteristics.

Case I – slab made of homogeneous material without defects. The temperature on side 2 uniformly increases during a transient till a steady state T_s is reached at time t_s which depends on the slab thickness and thermal diffusivity; then a uniform temperature is maintained with small variations in time (Figure 5).

Case II – a defect is buried inside the material. Again, two cases may arise depending on the thermal diffusivity of the defect αd with respect to that of the bulk sound material αb.

a) $\alpha d < \alpha b$ the defect acts as obstacle to the heat propagation. Then, the temperature distribution on side 2 during the transient will show a lower temperature value in correspondence of the defect (Figure 6a).

b) $\alpha d > \alpha b$ heat propagation is accelerated through the defected zone. Then, the temperature distribution on side 2 during the transient will show a temperature increase in correspondence of the defect (Figure 6b).

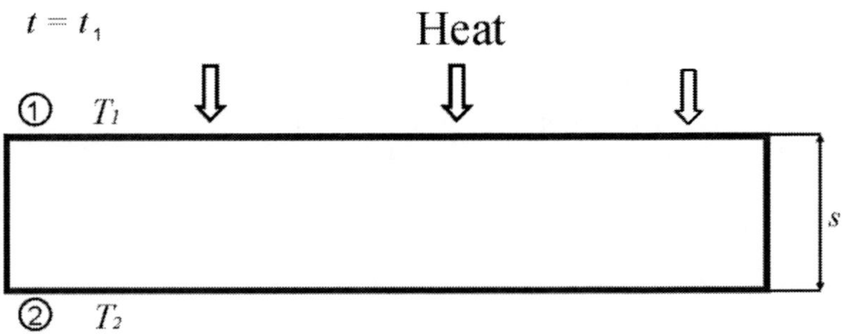

Figure 4. Heat transfer model.

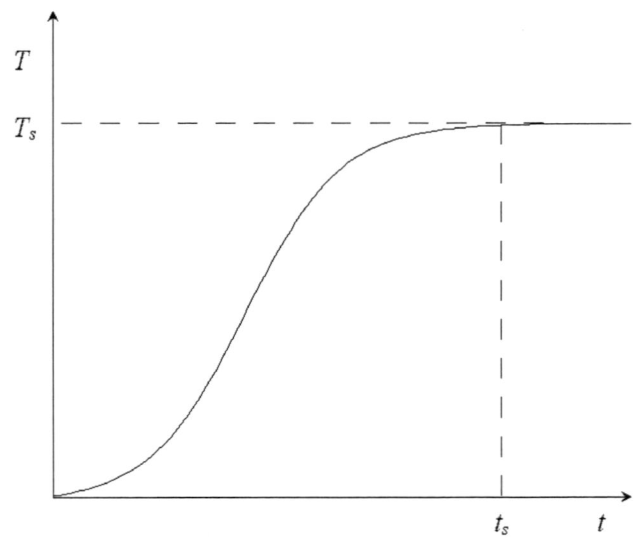

Figure 5. Temperature against time for a transient-steady state condition.

Defects can be identified as local temperature differences in the surface temperature map:

$$\Delta T = T_b - T_d \qquad (17)$$

with T_b the temperature in the sound material zone and T_d the temperature in the defected one. During transient, there is a time instant at which ΔT attains its maximum, i.e., defects are visualized with the best contrast. For a given slab, this time depends on defect position across the slab. Going towards steady state conditions, $t = t_s$, $\Delta T \to 0$ and it is no more possible to detect defects.

Therefore, the material can be simply evaluated by mapping the temperature variation over a surface under thermal stimulation. In fact, the propagation of thermal energy, within the material under inspection, can be modelled by the Fourier diffusion equation (16).

For uniform heating and in case of homogeneous material the surface temperature T_S distribution would be uniform. The presence of a discontinuity (defect) at a certain depth interferes with the propagation of the thermal energy causing a localised temperature difference $\Delta T = T_S - T_d$, the subscripts *s* and *d* respectively referring to a sound or to a defective area. Thus, defects are detectable as local discontinuities in the temperature map; but, such a map must be acquired during transient thermal stimulation.

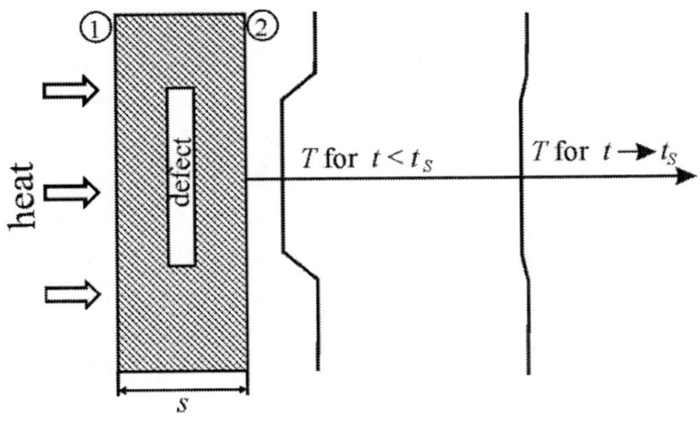

a) $\alpha d < \alpha b$

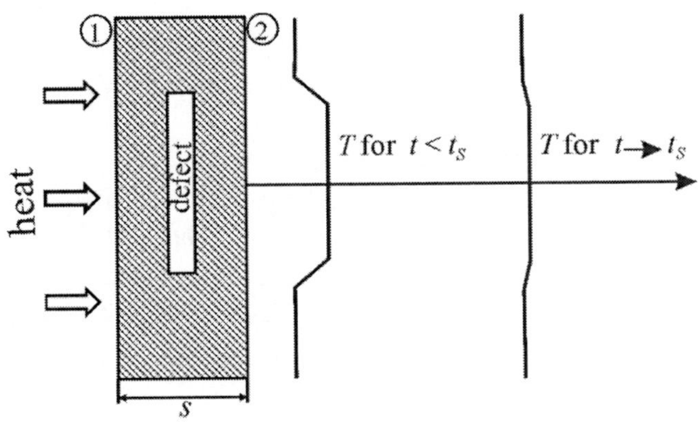

b) $\alpha d > \alpha b$

Figure 6. Temperature variation on side 2 for a) $\alpha d < \alpha b$ and b) $\alpha d > \alpha b$.

4. Thermographic Methods for Non Destructive Inspection

Basically, two thermographic techniques can be used for non destructive inspection: Pulse Thermography (PT) and Lockin Thermography (LT). A combination of both PT and LT gives the Pulse Phase Thermography (PPT) in which the specimen is pulse heated as in PT and the mix of frequencies of the thermal waves launched into the specimen is unscrambled by performing the Fourier transform of the temperature evolution over the field of view. In PPT data can be presented as amplitude, or phase, image like in LT. A comparison between these three techniques was performed by Carlomagno and Meola [46]. However, there are other techniques such as Time-Resolved-Infrared (TRIR), Vibrothermography, etc. which are variations of PT and LT because they include a different heating method, or a different processing algorithm. Of course, as technology evolves, new techniques may be tailored for a specific need.

4.1. Pulse Thermography

PT consists simply in the stimulation of the object (under evaluation) by a heating pulse and monitoring of its surface temperature variation during the transient heating, or cooling, phase. Heating is generally performed by lamps, flash lamps, scanning lasers, or hot air jets. Cooling can be practically attained mainly by cold air jets. Of course, air jets (hot or cold) can be used only on a massive surface since jet impingement may damage delicate artworks.

For the case of slabs, analysis with PT can be performed in two different modes (Figure 7): transmission and reflection. In transmission mode (Figure 7a), the infrared camera views the rear face, i.e. the face opposite to the heating/cooling source. But, since the opposite side is not always accessible and/or available, the reflection mode, (Figure 7b) for which both heating (cooling) source and camera are positioned on the same side, is mainly applied. The thermal energy propagates by conduction inside the material while the infrared camera monitors the temperature variation over the viewed surface. Obviously, for a uniform surface heating and homogeneous material, the temperature T distribution is uniform in case of a homogeneous material. The presence of a defect at a certain depth interferes with the heat flow causing local surface temperature variations. An example of temperature map for sound and defective materials is shown in Figure 8.

More specifically, both images refer to specimens made of glass fibres reinforced polymer material; the first image (Figure 8a) displays a sound material while the second one (Figure 8b) shows five defects of rectangular section. These images were taken with the Agema 900 LW camera equipped with a Mercury Cadmium Telluride (MCT) detector, liquid nitrogen cooled, working in the 8-12 μm band of the electromagnetic spectrum.

The visibility of the defect can be evaluated by the parameter D_T:

$$D_T = \frac{|\Delta T|}{|\Delta T_s|} = \frac{|T_s - T_d|}{|T_s - T_r|} \qquad (18)$$

where, as schematically shown in Figure 9, T_d is the temperature over a defective zone, T_s is the temperature in a sound zone and T_r is a reference temperature. More specifically, T_r is the temperature of the sound material before starting of transient heating, or cooling $T_r = T_{s(t=0)}$ (i.e. the temperature of the first thermal image taken at t = 0 s in the time sequence). Indeed, the quantity $T_s - T_r = \Delta T_s$ plays an important role because it indicates the optimal temperature variation to which the material has to be subjected for good defect visibility. The dimensionless temperature difference D_T in equation (18) can assume either positive, or negative values, the sign depending on the test procedure (acquisition during heating, or during cooling) and on the thermal properties of both the sound material and the material of which the defect is made.

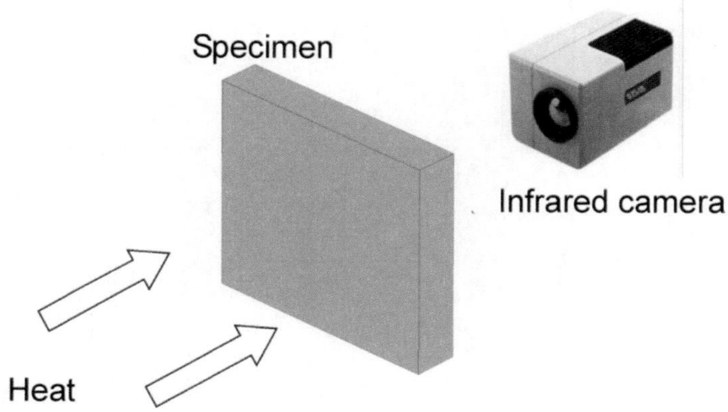

transmission mode

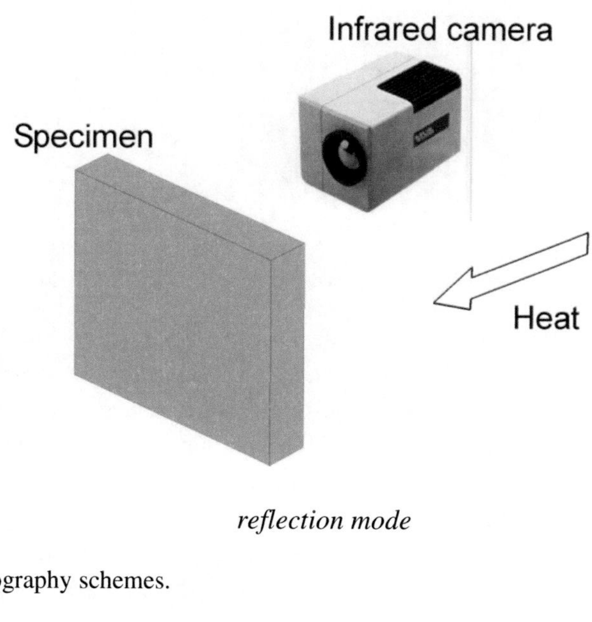

reflection mode

Figure 7. Pulse thermography schemes.

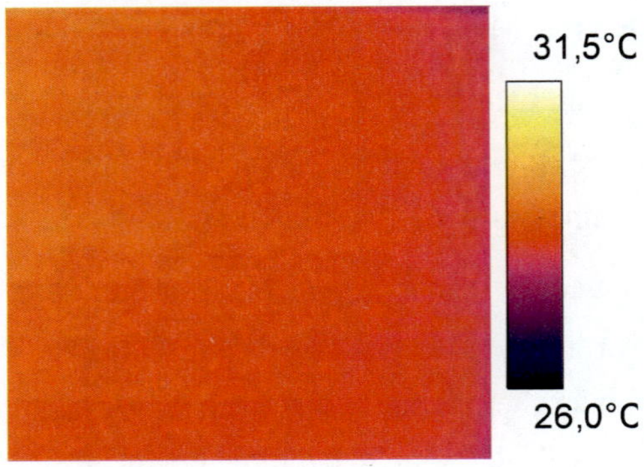

a) sound material

b) defective material

Figure 8. Temperature distribution for: a) sound material; b) material with defects

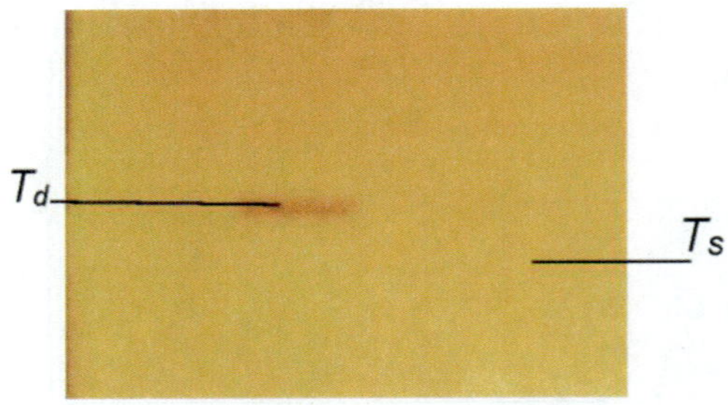

Figure 9. Schematic of temperature nomenclature over a defective specimen surface for Eq.(18)

Generally, a deep defect becomes visible later than a shallow one and a larger defect produces a more marked temperature difference. Thus, a complete (from surface to depth) evaluation of a material is performed by acquiring thermal images (also called thermograms) in a time sequence. The thermal image may be shown in grey, or colour levels on the computer monitor (10, 16,..., 112, or more), each level being related to a temperature interval; such interval depends on the selected range (the minimum allowable range for the scanner is often about 2 degrees). Information about the defect parameters such as size, depth and thermal resistance can be obtained by applying post-processing procedures to the thermograms.

The defect detection is limited by the Signal-to-Noise Ratio (SNR) [47], or by the Noise Equivalent Temperature Difference (NETD). In addition, the surface finish is of great importance since variations in surface roughness, cleanliness, uniformity of paint and other surface conditions can cause variations in the emissivity coefficient and affect the temperature measurement. These aspects were analysed by Meola and Carlomagno [1, 17].

4.2. Lock-in Thermography

The concept of lock-in thermography was first introduced by Carlomagno and Berardi [48] and later further investigated by other researchers [40-54]. Classical lock-in thermography is basically performed with a modulated heating lamp and is called optical lock-in thermography and abbreviated as LT, or OLT (some authors also call it modulated thermography MT). The energy, generated by halogen lamps, is delivered to the object surface in the form of periodic thermal waves. The thermographic system is coherently coupled to the thermal wave source which is operated in such a way that a sinusoidal temperature modulation results. This modulation is obtained from a non-linear electrical signal produced by the lock-in module which allows also for frequency variation. The system collects a series of images and compares the modulated heating to the measured temperatures by extracting the sinusoidal wave pattern at each point of the image. The harmonic heat transfer delivered at the surface (depth $z = 0$) of a homogeneous and semi-infinite material results in a (time dependent) thermal wave, which, in one dimension, is given by [45]:

$$T(z,t) = T_o \, exp\,(-\frac{z}{\mu})\, exp\, i\left(\omega t - \frac{z}{\mu}\right) = A(z)\ exp\, i\left[\omega t - \phi(z)\right] \tag{19}$$

where $A(z)$ is the thermal amplitude, $\phi\,(z)$ is the phase shift of the thermal wave travelling inside the material, μ is the thermal diffusion length which is calculated from the thermal diffusivity α and the wave frequency $f = \omega / 2\pi$:

$$\mu = \sqrt{\frac{\alpha}{\pi f}} \tag{20}$$

In the classical optical lock-in method both lamp and camera are located on the same specimen side (Figure 10).

The thermal wave propagates inside the material and gets reflected when it reaches parts where the heat propagation parameters change (in-homogeneities). The reflected wave interferes with the surface wave producing an oscillating interference pattern, which can be measured in terms of amplitude $A(z)$, or phase angle $\phi\,(z)$, that, respectively, produces amplitude, or phase, images. The depth range for the amplitude image is given by μ while the maximum depth p, which can be reached for the phase image, corresponds to 1.8μ [55-58].

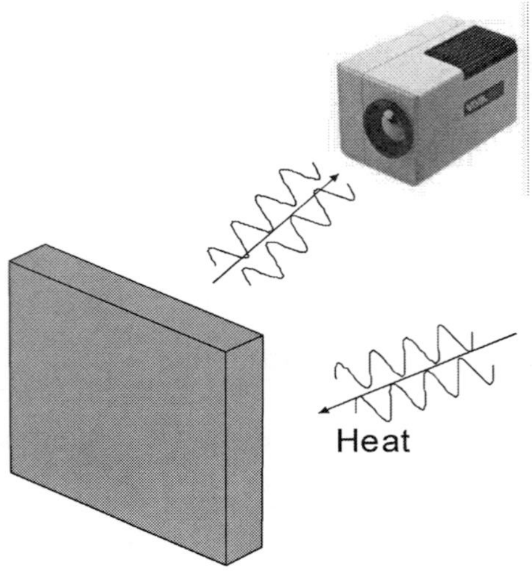

Figure 10. Lock-in thermography scheme.

Therefore, the material thickness, which can be inspected, depends on the wave period (the longer the period, the deeper the penetration) and on the thermal material properties (thermal conductivity and volumetric specific heat). Generally, tests start at a quite high wave frequency, at which only surface (or low depth) defects are visible, and later on, to inspect deeper layers, the frequency is decreased until the entire thickness has been traversed, or the minimum selectable frequency has been reached. Results are generally presented as phase images where a local variation of colour indicates a local variation of phase angle ϕ and, in turn, a local variation of material thermal properties. The images may be stored for successive analysis to acquire information about size, depth and nature of defects.

As an example of material evaluation with OLT, the specimen sketched in Figure 11 is considered; this specimen is made of plastic 8 mm thick and includes several back drilled holes of different diameter D = 4, 6, 8, 10 mm and at different depth p in the range 1.4 up to 6 mm from the flat surface. The evaluation was performed with the Flir Systems ThermaCam SC3000 Quantum Well Infrared Photonic detector (QWIP) Stirling cooled working in the 8-9 μm band of the electromagnetic spectrum. For LT purposes this camera is equipped with the IRLockIn option. Four phase images, which were taken at different heating stimulation frequency values, are reported in Figure 12. More specifically, the first image (Figure 12a), which was taken at f = 1.5 Hz, displays an almost homogeneous material surface condition. At the lower frequency of f = 0.5 Hz (Figure 12b) the shallow defects appears; at a much frequency decrease, more defects are detected (Figure 12c) and, at a further frequency reduction (Figure 12d), practically all the defects are visualized.

By comparing Figs. 12b, 12c and 12d, it is possible to note that for f = 0.05 Hz all defects become visible but with a change of colour (i.e., change of phase angle). This is an important phenomenon to be taken into account when performing OLT tests, which will be discussed later in the next paragraph.

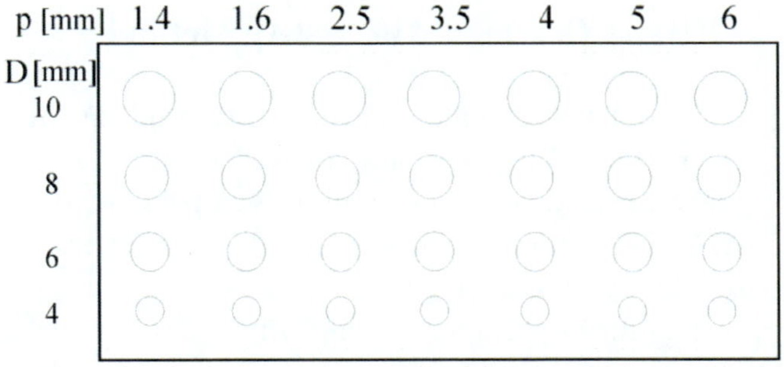

Figure 11. Sketch of the plastic specimen with back drilled holes.

a) $f = 1.5$ Hz

b) $f = 0.5$ Hz

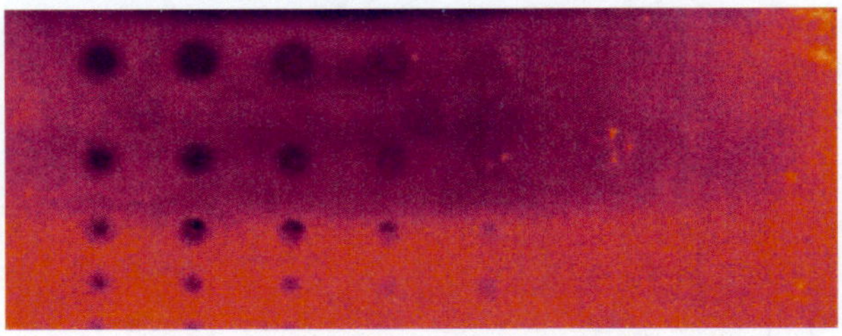

c) $f = 0.25$ Hz

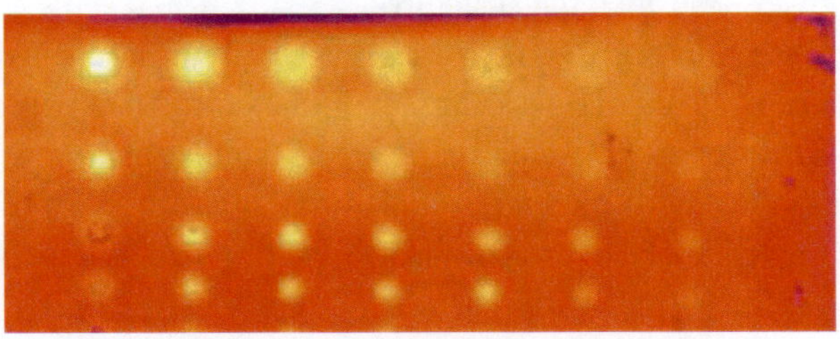

d) $f = 0.05$ Hz

Figure 12. Example of defects detection with varying heating modulation frequency.

4.3. Limitations of PT and OLT and Alternative Methods

One main requirement in PT is the uniform distribution of the heat flux over the surface of the object under examination. This is simply attained in the case of a relatively small surface, while it becomes more difficult for large surfaces. A possible solution may be the use of an array of halogen lamps. For some applications, it is convenient the *Lateral Heating* (LHT) method conceived by Grinzato et al. [59].

In specific cases such as the inspection of cavities and channels, when a strong heat pulse is required, the pulse thermography with *Injecting Water Vapour* (PTJV) may be convenient. PTJV was proposed by Meola et al. [60] to detect ceramic residuals in the tip of turbine blades. The advantage of using steam is that vapour in contact with a cold wall results in a very high convective heat transfer coefficient so that the wall surface contacting the vapour gets suddenly very close to the vapour condensing temperature, thus determining a strong heat pulse.

A technique which has some advantages over pulse thermography is the *Time-Resolved-Infrared* (TRIR) technique described by Aamodt et al. [61]. Within this technique a step heating (long pulse) is applied to the specimen while its surface temperature is monitored as a function of time. The heating pulse is long compared to the observation time and acts as a temporal step function; this technique is particularly indicated for the analysis of multilayered coatings. Spicer and colleagues (1999) have also introduced *microwave heating* into the TRIR technique and found this method advantageous to detect subsurface absorbing microwave features such as water-filled areas, metal wires or fibers if the host is transparent to microwaves.

It has to be taken in mind that LT performed with optical heating is sensitive to thermal boundaries, which do not allow for a clear and reliable discrimination of the intact areas from those of mechanical weakness. For the case of gas-filled defects, such as cracks, or voids, this problem may be solved by selectively stimulating defective areas with acoustic waves. This is an alternative deployment of lockin thermography, which is called *Ultrasound Lockin Thermography* (ULT) [22]. More specifically, the object is stimulated with acoustic waves which permit a selective heating of only defective areas; such technique is particularly indicated for detection of corrosion, vertical cracks and delaminations. More generally, infrared thermography associated with the detection of thermal energy produced by mechanical vibration is called *vibrothermography*. Under mechanical vibration, heat is released by friction where discontinuities such cracks and delaminations are located [62]. Vibrothermography [63] is also known as SPATE (*Stress Patterns Analysis by the measurement of Thermal Emission*), which is a tool to obtain full-field non-contact stress data and is particularly useful for metallic structures.

However, the detection of solid inclusions, especially if the inclusion has properties close to component material, is a crucial task. In fact, in the basic OLT approach the reflected wave is exploited for defect detection. Actually, a part of the thermal wave gets reflected back W_r, another part diffuses W_d, in all directions, and another part propagates W_p inside the material and reaches the opposite slab surface. In presence of an anomaly, each of these three components may modify in a different way depending on depth and properties of the anomaly. In particular, some shallow defects are well detected in the reflected way (lamp and camera on the same side). Instead, the detection of a defect with OLT becomes difficult if such defects is buried deep and is a thin small insert of thermal properties close to the hosting material [64]. If otherwise the hosting materials has complex geometry (e.g., curved) and bright surface, the defect detection becomes nearly impossible. In fact, in this case, the reflected signal W_r becomes too weak due to the diffusion inside the sound material and is hindered by the strong reflection of the lamp on the bright curved surface. One would infer that the phase image is not affected by reflections from the surrounding, but this is true to the extent that such reflections remain constant with time. In some cases the magnitude of the reflections may change as the surface becomes hot.

This drawback can be overcome with the *Rear Heating Phase Thermography* approach (RHPT) where the transmitted part of the thermal wave is exploited for defects detection [65]. Thin inclusions are resolved better than with the basic OLT; a great advantage of RHPT is the possibility for close-up view inspection since short focal distances do not permit positioning of lamps on the front-side.

Another technique which combines the advantages of both pulse and lockin thermography without sharing their drawbacks is the *Pulse Phase Thermography* (PPT), which was introduced by Maldague and Marinetti in 1996 [66] and later analysed by Vavilov and Marinetti (1999) [67]. Within PPT, the specimen is pulse heated as in PT and the mix of frequencies of the thermal waves launched into the specimen is unscrambled by performing the Fourier transform of the temperature evolution over the field of view; the phase, or magnitude, image can be presented as in LT. Deeper thicknesses can be inspected with less influence of surface infrared and optical features. A comparison between PT, LT and PPT applied to the evaluation of frescoes was presented by Carlomagno and Meola (2002) [46].

5. Defects Characterization

Once the most effective technique is chosen, the evaluation of large objects should be made by first taking an image of their entire surface; this permits detection of the larger defects. Then, for a detailed analysis, the viewed surface should be partitioned into several zones to be separately tested. For defects location (over the full surface), it is convenient to place markers on the corners of each zone; indeed, the coordinates of each defect can be precisely evaluated by knowing the viewing distance and the focal length of the employed lens. Then, the entire surface of the inspected object may be reconstructed by patching up all the stored images (one for each zone).

The geometry and location of defects can be evaluated from the thermal, or phase, images. The size of defects may be measured from the images by considering the local temperature (or phase angle) variation from sound to damaged zones and obviously by exactly knowing the spatial resolution of the employed lens. Depending on the thermal properties of the damaged zone with respect to the sound material, temperature or phase angle differences may display positive, or negative, values.

The main problem in the measurement of a defect size, especially in thermal images (obtained with PT), depends on the thermal diffusivity within the material. For $\alpha d < \alpha b$ (sect.3.1), the temperature difference ΔT over a damaged zone with respect to the sound material attains its maximum value over the centre of the defect and decreases laterally going towards the sound material in a "V" shaped way as sketched in Figure 13.

Therefore, the boundary between the defect and sound material is not clearly identified and worsens with increasing defect depth. A criterion may be to establish the minimum Δ value to discriminate between sound material and defect by considering both the heat transfer mechanisms and the noise signal [1, 8, 60].

By introducing the parameter ψ to indicate either the temperature ($\psi = T$), or the phase angle ($\psi = \phi$), value, depending on the image under analysis (thermal image, or phase image), and by assuming ψ_m as the average value of such parameter for sound material, and ψ_c as the value on the surface above the centre of the defect, it is possible to define the defect contour as the locus of values ψ which satisfy the following conditions:

$$\frac{\psi_m - \psi}{\psi_m - \psi_c} \cong 0.5$$
$$\psi_m - \psi > \Delta\psi_s \tag{21}$$

where $\Delta\psi_s$ is the average deviation of ψ over the sound material. Besides the dimension, also the nature of defects is of utmost importance. Methods, based on inversion procedures, were also developed to understand the severity of the damaged area, but such methods were not at all successful as they were time consuming and limited to specific cases. Thus, simplified analytical approaches based on dedicated measurement of thermophysical properties are generally preferred.

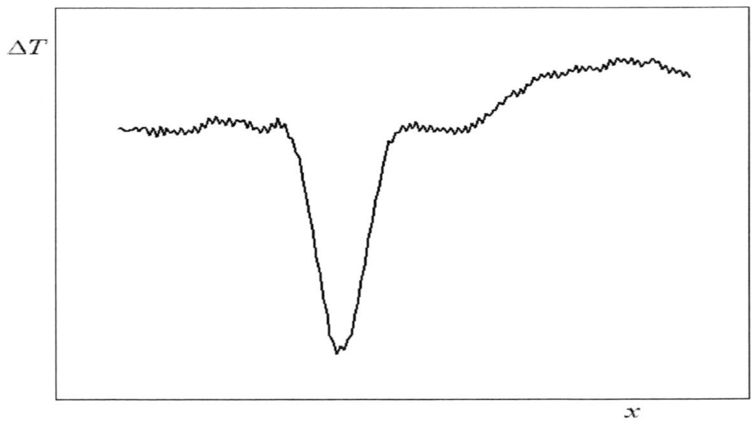

Figure 13. Temperature spatial variation above a defect.

In this context, the LT technique may be preferable since it allows for simple and fast defects characterization without troublesome post-processing procedures. In fact, the phase angle can be assumed as a characteristic material parameter [65] and Eq.(20) supplies information about defect depth and thickness [1, 17]. The LT technique itself can be used, following the procedure described in Ref. [11], to evaluate the thermal diffusivity α itself. More specifically, for the material that is under exam, by considering a homogeneous specimen (without defects) of known thickness s_p, one has to find the frequency f_p for which the entire thickness s_p is traversed. By introducing the two values f_p and s_p in equation (20), which is rearranged in the form:

$$\alpha = \pi f_p \left(\frac{s_p}{1.8} \right)^2 \tag{22}$$

with $s_p/1.8 = \mu$, the searched α value is obtained. This is a great advantage because the overall thermal diffusivity of composites is not always available since it is influenced by several factors such as the type of materials (fiber and matrix), the volume of fibers with respect to that of the matrix, the fibers orientation and stacking sequence.

With regard to the defect thickness, one has to consider Eq. (21) rearranged in terms of phase angle contrast ($\psi = \phi$):

$$C_\phi = \frac{\phi m - \phi}{\phi m - \phi c} \tag{23}$$

It has to be taken in mind that $\Delta\phi$ may change sign with varying the wave frequency ($\Delta\phi$ may become positive, zero, or negative) [1]. More specifically, when the thermal wave (of suitable frequency value) touches the defect top surface, the defect appears, but with poor contrast (only a mark) and $\Delta\phi$ could be positive, or negative. Generally, decreasing the frequency, the contrast enhances and becomes the highest as the thermal wave approaches the defect centre across its thickness. Then, it worsens again, vanishing ($\Delta\phi$ = 0) when the thermal wave reaches the defect bottom; the corresponding frequency is called *blind frequency* f_b. After that, the defect appears again but with a change of colour (change of $\Delta\phi$ sign). A typical variation of $\Delta\phi$, with an initial negative value, with varying the thermal wave frequency is represented in Figure14.

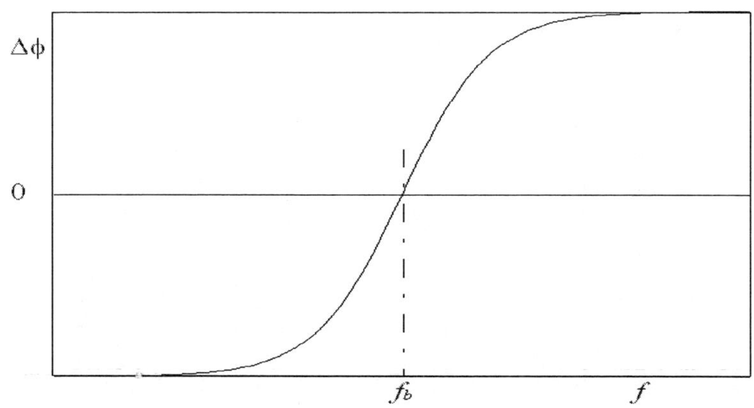

Figure 14. $\Delta\phi$ distribution with f.

6. APPLICATIONS

By choosing the most suitable method, infrared thermography is capable of detecting the majority of the recurrent types of defects in many materials and structures. This section shows some examples of NDE with IRT which were performed by the research group of the University of Naples. Indeed, this research group has been working with infrared thermography for long time and investigated IRT capability in several different fields such as fluid-dynamics, materials NDE, architecture and cultural heritage [1-2, 11, 46, 48, 60, 64-72].

In particular, within the materials NDE, the investigation has been driven towards the suitability of IRT to deal with different types of materials problems [1-2, 8, 11, 13-14, 15, 17, 60, 64-68] and to search for the most effective technique and/or procedure for a given application [12, 48, 64, 69]. Few examples of materials evaluation are reported and discussed in the next section. To get a more exhaustive *panorama* on NDE with IRT, the reader is addressed to literature. For convenience, some of the main works, which were produced by the many research groups through the world, are collected in Table 1.

Amongst the examples herein shown, particular attention is devoted to the inspection of fibre reinforced composite materials due to the great interest they rouse in the scientific and industrial community. This because of the several advantages they offer over metals [73-74].

Table 1. A collection of applications

Reference	Material	Objective	Technique	Notes
Tenek and Henneke, 1991 [62]	Composites	Flaws detection	Vibrothermography	
Karpen et al., 1994 [52]	Composite	Fibre orientation	LT	
Wu, 1994 [53]	Wood	Delamination detection	LT	
Rantala et al., 1996 [63]	Polymers, composites	Flaw detection	Vibrothermography	
Dillenz et al., 2003 [22]	Metals	Vertical cracks	ULT	
Meola et al. [13]	Bonded structures	Various	PT, OLT	
Meola et al. 2005 [66]	Composites	Slag inclusions, delamination	OLT	
Carlomagno and Meola, 2002 [46]	Fresco	Detachment of tiles, cracks in the support, etc.	PT, OLT, PPT comparison	
Meola et al., 2003 [14]	Glare®	Behaviour under bearing load	OLT	

Table 1. (Continued)

Reference	Material	Objective	Technique	Notes
Meola et al., 2004 [15]	Polyethylene	Crosslinking distribution	OLT	
Thomas et al., 1980 [58]	Metal	Flaw detection	LT	
Busse, 1979 [57]	Metals	Flaw detection	LT	
Grinzato et al. 1998 [59]	Frescoes		LHT	
Aamodt et al., 1990 [61]		Multilayered coatings	TRIR	
Carlomagno et al., 2005 [67]	Masonry structures	Voids, cracks, etc.	PT, OLT	Comparison with geo-electric methods and ultrasounds
Meola, 2007 [68]	Masonry structures	Voids, cracks, water infiltrations, etc.	PT	Data correlation
Meola et al., 2008 [60]	Metals	Ceramic residuals in turbine blades	PT, PTV, OLT	Comparison with X rays
Meola et al., 2008 [71]	CFRP	Impact damage	OLT	Comparison with SQUID
Meola, 2007 [69]	Various	Thin inclusions	RHPT	
Meola et al., 2009 [72]	GFRP	Impact damage	OLT	

6.1. Composite Materials

Their greatest advantage is strength and stiffness combined with lightness, which makes such materials attractive for the aircraft industry. Another important advantage is the crack-stopper aptitude of the reinforcement; in fact, an important primary action of the fibres is to impede the widening of a small crack and to share the stress around. The same crack in a metal would rapidly enlarge compromising the part's performance. With the appropriate choice of constituents, composite materials represent a durable alternative in extreme environments to overcome the corrosion problem. A not least important aspect is the design flexibility; in fact, the material can be designed to meet the desired performance. In addition, composites can be easily moulded into complex shapes. The sum of all these advantages has been the force driving the development of composites.

However, in contrast with the benefits previously mentioned, composites also pose a strong challenge by analysis and evaluation techniques and procedures, which also require continue upgrading to keep up with materials innovation. Thus, it is justified the renewed interest on nondestructive evaluation (NDE) technology is justified. Indeed, the need of ever more sophisticated NDE techniques encompasses two main motivations. On one side, some of the techniques which are well assessed for metals, may be not effective for novel materials. On the other side, there is an increasing demand of *quality* and *safety* especially in the aeronautical field. High quality standards require the entire production being *defects free* and this can be only assured through NDE of each component and of the whole structure by using the most effective nondestructive technique.

Three types of defects are mainly encountered in composites which are slag inclusions, impact damage and delaminations, which, in the following, will be separately treated .

Slag inclusions. A carbon fibre reinforced polymer (CFRP) is considered with thin Teflon inserts of different dimensions and placed at different depths. This would simulate the inclusion of the backing film, which is harmful for the component's life because it acts as initiator site once the component becomes a load-bearing structure.

The specimen was fabricated by superimposing several epoxy resin pre-impregnated carbon fabric $(0,90)_f$ layers to have a step wedged profile as sketched in Figure15 (not to scale) and allowed to cure in a press at ambient temperature. The entire specimen length is divided into 9 steps of thickness varying from 2.2 up to 7 mm, each step being 50 mm wide and 175 mm high. Two Teflon square inserts of thickness s_T = 200 μm, surface area equal to either 40 or 160 mm^2, are positioned at half thickness depth in each step so as to have a defect depth going from p = 1.1 up to 3.5 mm.

Tests were performed with a focal plane array (FPA) Flir SC3000 QWIP camera and with the Rear Heating Phase Thermography [69]. A phase image taken at f = 0.5 Hz is shown in Figure16 and as can be seen, at this frequency, all inclusions at every depth are visible even if [65] the s_T/p ratio is very small.

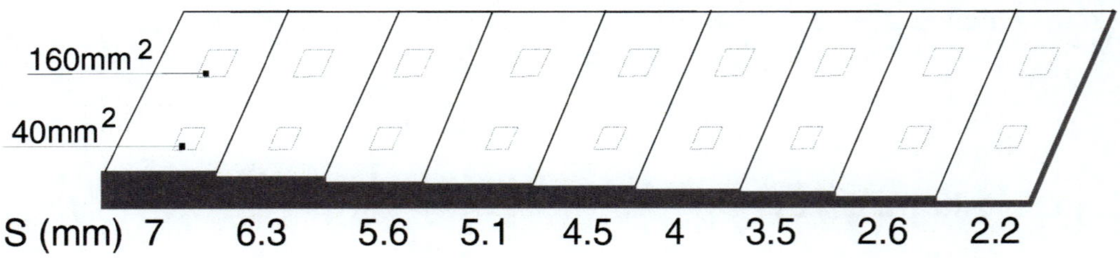

Figure 15. Sketch of CFRP specimen with slag inclusions.

Figure 16. Phase image of the CFRP specimen with slag inclusions.

Impact damage. The main weakness of composites is their vulnerability to impact damage. In fact, a relatively low velocity impact may not produce perforation but just delamination between the layers with no visible surface evidence. Such a delamination may grow in service causing severe stiffness reduction in the structure, which may lead to catastrophic failure. Due to the high probability of impact-damaging composite materials during manufacture, service and maintenance, periodic inspection with adequate predictive methods is required. On the other side, impact damage in composites is a very complex mechanism that involves matrix cracking, surface buckling, delamination, fibre shear-out and fibre rupture [75-79].

Three specimens of different materials are considered for detection of impact damage.

1) The first specimen is a glass reinforced polymer (GFRP) 2 mm thick, subjected to three impacts of different energy, respectively 1.2, 1.3 and 2.1 J.
2) The second specimen is a carbon reinforced polymer (CFRP) 3 mm thick with three impacts of 2.2, 2.3 and 3.1 J.
3) The third specimen is a sandwich CFRP/ Nomex® with one impact of 5 J. At the front side of the latter specimen a painted strip over the CFRP surface to simulate the presence of a local different material is present. At the backside, a Nomex® honeycomb panel, which is tapered in part to have thickness variation, is present.

All the specimens are tested with the FPA Flir SC3000 QWIP camera with optical lock-in thermography; the camera views the impacted side.

Two superimposed phase images of both GFRP and CFRP specimens are shown in Figure 17. The impact energy on the GFRP specimen (above) increases from left to right. The CFRP specimen (below) has impacts of energy increasing from right to left.At the frequency of 0.5 Hz (Figure 17a) only the strongest impact of 2.1 J is visible on the GFRP specimen while, it is possible to distinguish all the three impacts on the CFRP specimen. As the frequency is decreased to 0.25 Hz (Figure 17b) the impacted areas appear better outlined; from these images it is also possible to gain information about the extension of the damage around the impact point (darker halo around each impact in the CFRP specimen). To evidence the lowest energy impacts in GFRP, it is necessary to further decrease the heating frequency.

A comparison between the phase images of the two specimens supplies also information about the different behaviour of the two composites when impacted. It seems that the damage is close to the impact point in the GFRP, while it extends to a certain distance in the CFRP.

A phase image, taken at 0.5 Hz, of the CFRP/ Nomex® sandwich specimen is reported in Figure 18. As can be seen, it is possible to distinguish not only the low-energy impact but also the thickness variation of the specimen. In fact, the camera viewed the impacted side where, apart from the copper colour stripe, the surface appeared uniform without any thickness variation since the tapering is on the opposite side. By comparing this image (Figure 18) with the image of the CFRP specimen (Figure 17), it is also possible to get information about the behaviour of the honeycomb in an impacted sandwich.

Infrared thermography has proved useful also for gaining information about the impact dynamics and the material behavior under impact load. These pieces of information can be obtained by monitoring with an infrared camera the thermal response of composites during an impact event [72, 80]. What is important is the choice of the infrared radiometer that should be equipped with a high resolution (both thermal and spatial) detector and allow for adequate frame rate.

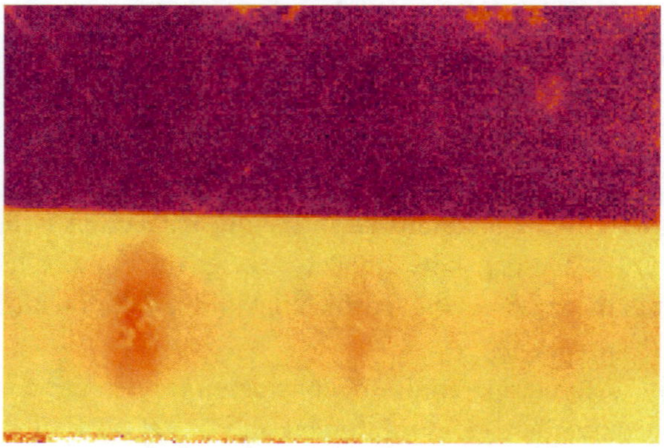

a) $f = 0.5$ Hz

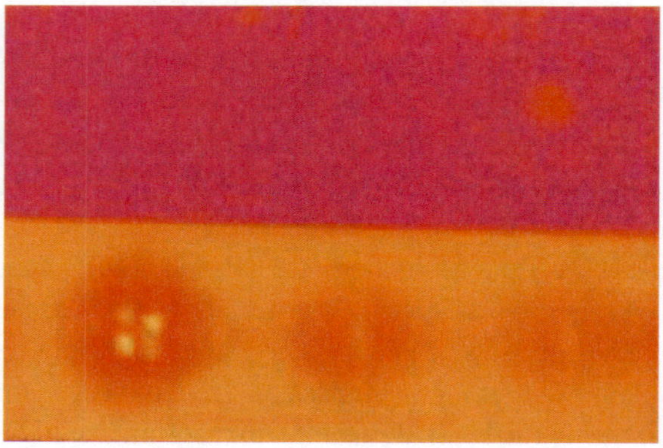

b) $f = 0.25$ Hz

Figure 17. Phase images of both GFRP (above) and CFRP (below) impacted specimens.

Figure 18. Phase image of CFRP/Nomex® sandwich specimen taken at $f = 0.5$ Hz.

Pin-bearing load. The last example regards the evaluation of the damage under pin-bearing load of a Glare® specimen. Glare® belongs to the family of materials called Fibre Reinforced Metal Laminates (FRML), which include thin layers of metal sheet and unidirectional fibre layers embedded in an adhesive system. This material originates from studies performed in the early seventies at the Delft University of Technology. At that time the efforts to create a material with improved fatigue properties and high damage tolerance capabilities were driven by the Boeing 737 accident in which, after intensive use, many fatigue cracks occurred and a large part of the fuselage separated in flight from the aircraft. This hybrid composite material includes thin aluminium Al 7075 T-7351 layers combined with R glass fiber/epoxy layers.

Two pieces 20 mm wide and 50 mm long were cut from a large laminate and a hole of diameter D = 8 mm was drilled in each. One of the specimens was subjected to bearing failure. In Figure 19, the phase images of both the unloaded specimen (Figure 19a) and the failed one (Figure 19b) are shown. A light arc on the hole right side noticeably indicates the damage caused by pin-bearing load.

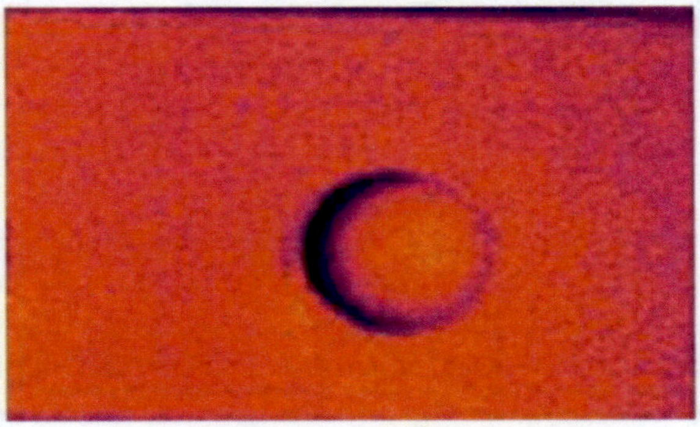

a) unload

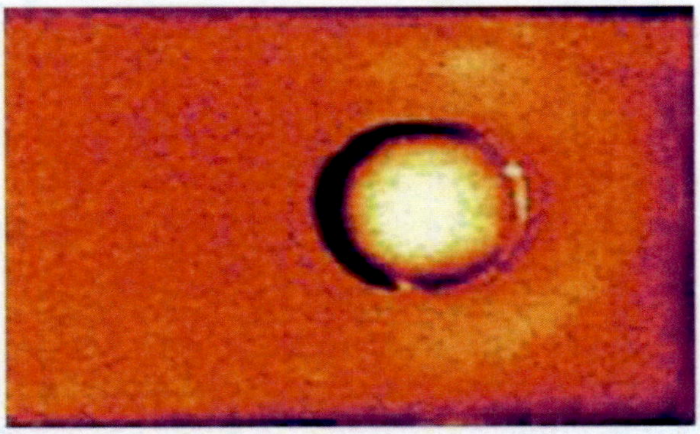

b) after bearing failure

Figure19. Phase images of the Glare® specimen.

CONCLUSION

The use of infrared thermography in non destructive inspection of materials has been reviewed. Some basic concepts were recalled and the methods mainly used for non destructive testing and evaluation were discussed and some applications were presented.

Some important points are herein recalled:

- Infrared thermography, as a remote imaging technique, can be exploited for non destructive evaluation of parts both during manufacturing and in service.
- Pulse thermography (PT) is the simplest infrared thermography technique, it can be performed with almost any infrared camera, but it suffers from non uniform thermal stimulation of the surface and emissivity variation unless it is used as pulse phase thermography (PPT) which is practically equivalent to lock-in thermography.
- Lock-in thermography (LT) requires an additional component to modulate the stimulation (heating) source, but it is not affected by surface emissivity variation and non uniform thermal stimulation. In addition, LT allows for a quite quick evaluation of the depth at which a defect is located.

On the basis of the obtained results, and of also those not shown herein, it is possible to infer that pulse thermography is mainly capable of detecting inclusions of foreign materials, formation of air bubbles, impact damage disbondings and delaminations. Besides, lock-in thermography is also capable of supplying information about the adhesive thickness, the paint thickness, the fibres stacking sequence and the influence of substrates surface treatments on adhesion.

The main advantages of the infrared thermography NDT technique are: non-contact, non-intrusive, two-dimensionality, remote imaging of large areas. Notwithstanding this, infrared thermography is still not completely exploited since it is still not adequately enclosed in industrial instrumentation. Conversely, IRT can be used for lots of applications; what is important is the choice of the most adequate instrument and technique.

REFERENCES

[1] Meola, C.; Carlomagno, G.M. *Meas. Sci.Technol.* 2004, *15*, 27-58.
[2] Meola, C.; Giorleo, G.; Nele, L.; Squillace, A.; Carlomagno, G.M. *Nondestr. Testing Eval.* 2002, *18*, 83-90.
[3] Proc. *Annual SPIE Thermosense Conferences*, Bellingham, WA 1978–
[4] Proc. *Advanced Infrared Technology and Applications (AITA)* 1991-.
[5] Proc. *Biennial Quantitative Infrared Thermography (QIRT) Conference*s, 1992–.
[6] Cielo, P.; Maldague, X.; Déom, A.A.; Lewak, R. *Mat. Eval.*, 1987, *45*, 452-460.
[7] Vavilov, V. in *Proc. QIRT92;* Editors, D. Balageas, G. Busse and G.M. Carlomagno; EETI editions Paris, France, 1992; pp. 179-193.
[8] Giorleo, G.; Meola, C. *ASM Int. J. Materials Engineering and Performance* 1998, *7*, 367-374.
[9] Karpen, W.; Wu, D.; Steegmuller, R.; Busse, G. in Proc. *QIRT 94* Editors, D. Balageas, G. Busse and G.M. Carlomagno; EETI editions Paris, France, 1994; pp. 281-286.
[10] Wu, D.; in Proc. *QIRT 94* Editors, D. Balageas, G. Busse and G.M. Carlomagno; EETI editions Paris, France, 1994; pp. 298-302.
[11] Meola, C.; Carlomagno, G.M.; Squillace, A.; Giorleo, G. *Meas. Sci. Technol.,*2002, *13,* 1583-1590.
[12] Carlomagno, G.M.; Meola, C. *NDTandE International,* 2002, *35*, 559-565.

[13] Meola, C.; Carlomagno, G.M.; Giorleo, L. *J. Adhesion Sci. Technol.*, 2003, *17*, 1207-1222.

[14] Meola, C.; Squillace, A.; Giorleo, G.; Nele, L. *J. Composite Materials,* 2003, *37*, 1543-1552.

[15] Meola, C.; Carlomagno, G.M.; Prisco, U.; Vitiello, A. *Research in Nondest. Eval.,*2004, *15*, 55-63.

[16] Maldague, X. *Theory and practice of infrared technology for non-destructive testing.* John Wiley and sons, inc., 2001.

[17] Meola, C.; Carlomagno, G.M.; *J. Adhesion Sci. Technol.,* 2006, *20*, 589-632.

[18] Beller, W.S. *Missiles and Rockets,* 1965, *16.*

[19] Alzolof, F.E. *Mater. Eval.* 1965, *11*, 234-242.

[20] Green, D.R. *Appl. Opt.* 1968, *7*, 1779-1786.

[21] Merhib, S.; Taylor, R. Techn. Rep. AD-740654, Watertown, MA 1972, 1972.

[22] Dillenz, A.; Zweschper, T.; Riegert, G.; Busse, G. *Rev. Sci. Instr.* 2003, *74*, 417-419.

[23] Soloman, S. Sensors Handbook McGraw Hill Professionals 1998.

[24] Rogalski, A. *Report on Progress in Physics*, 2005, *68*, 2267-2336.

[25] Wachter, E.A.; Thundat, T.; Oden, P.I.; Warmack, R.J.; Datskos, P.G.; Sharp, S.L. *Rev. Sci. Instrument* 1996, *67*, 3434-3439.

[26] Levinstein, H.; Mudar, J. in Proc. *IEEE*, 1975, vol.63, pp. 6-14.

[27] Kreisler, A.J.; Gaugue, A. *Supercond. Sci. Technol.* 2000, *13*, 1235-1245.

[28] Rogalski, A. *Progress in quantum electronics*, 2003, *27*, 59-210.

[29] Konstantatos, G.; Howard, I.; Fischer, A.; Hoogland, S.; Clifford, J.; Klem, E. *Nature Letters*, 2006, *13,*442.

[30] Armani, N.; Ferrari, C.; Salviati, G.; Bissoli, F.; Zha, M.; Zappettini, A.; Zanotti, L. *J. Physics Condensed Matter*, 2002, *14*, 13203-13209.

[31] Tribolet, P.; Vuillermet, M.; Destefanis, G. in Proc. SPIE *Advances in Optical Thin Films II*, C. Amra, N. Kaiser and A.H. Macleod Ed., 2005, vol. 5964, pp.49-60.

[32] Cabanski, W.; Munzberg, M.; Rode, W.; Wendler, J.; Ziegler, J.; Fleissner, J.; Fuchs F.; Rehm, R.; Schmitz, J.; Schneider, H.; Walther, M. in Proc. *Infrared Technology and Applications XXXI* B.F. Andresen, G.F. Fulop Ed. 2005, vol. 5783, pp. 340-349.

[33] Sizov, F.F. *Semiconductor Physics, Quantum Electronics and Optoelectronics*, 2000, *3*, 52-58.

[34] Janesick, J.; Putman, G. *Annual Review of Nuclear and Particle Science*, 2003, *53*, 263-300.

[35] Rose, A. Concepts in photoconductivity and allied problems, *Interscience*, New York 1963.

[36] Rogalski, A. *Opto-Electron. Rev.* 2006, *14*, 87-101.

[37] Norton, P.; Campbell, J.; Horn, S.; Reago, D. in *Proc. SPIE,* 2000, vol. 4130, pp. 226-236.

[38] Tribolet, P.; Destefanis, G. in Proc. *SPIE*, 2005, vol. 5783, pp. 37-53.

[39] Rogalski, A.; Martyniuk, P. *Infrared Physics and Technology*, 2006, *48*, 39-52.

[40] Gunapala, S.D.; Bandara, S.V.; Liu, J.K.; Hill, C.J.; Rafol, S.B.; Mugolo, J.M.; Trinh, J.T.; Tidrow, M.Z.; Le Van, P.D. *Semiconductor Sci. Technol.* 2005, *20*, 473-480.

[41] Nesher, O.; Klipstein, P.C., *Opto-Electron. Rev*. 2006, *14*, 61-70.

[42] Rehm, R.; Walther, M.; Schmitz, J.; Fleiβner, J.; Fuchs, F.; Ziegler, J.; Cabanski, W. *Opto-Electron. Rev*. 2006, *14*, 19-24.

[43] Handbook of *Infra-red detection technologies* edited by M. Henini and M. Razeghi, Elsevier 2002.
[44] Carslaw, H.S.; Jaeger, J.C. *Conduction of Heat in Solids*, Clarendon Press, Oxford, UK, 1959.
[45] Cattaneo, C. *Comptes Rendues*, 1958, *247*, 431-433.
[46] Carlomagno, G.M.; Meola, C. *NDTandE International*, 2002, *35*, 559-565.
[47] Vavilov, V.P.; Almond, D.P.; Busse, G.; Grinzato, E.; Krapez, J.-C.; Maldaque, X.; Marinetti, S.; Peng, W.; Shirayev, V.; Wu, D. in: *Proc. QIRT 98*, D. Balageas, G. Busse and G.M. Carlomagno (Eds), Akademickie Centrum Graficzno-Marketingowe, Łódź, Poland, 1998, pp. 43–52.
[48] Carlomagno, G.M.; Berardi,P.G. in Non-Destructive Testing, Proc. *III Infrared Information Exchange*, Warren C. ed., St. Louis, 1976, pp. 33-40.
[49] Busse, G. *Appl. Opt.* 1982, *21*, 107-110.
[50] Beaudoin, J.L.; Merienne, E.; Danjoux, R.; Egee, M. in Proc. SPIE *Infrared technology and applications,* 1985, *vol.* 590, pp. 287-292.
[51] Kuo, P.K.; Feng, Z.J.; Ahmed, T.; Favro, L.D.; Thomas, R.L.; Hartikainen, J. in Proc. *Photoacoustic and photothermal phenomena*, P. Hess, J. Pelzl (Eds.), Springer-Verlag, Heidelberg, 1987, pp. 415-418.
[52] Karpen, W.; Wu, D.; Steegmuller, R.; Busse, in Proc. *Qirt 94, Eurotherm Series 42,* D. Balageas, G. Busse and G.M. Carlomagno Eds. EETI editions, Paris, France, 1994, pp. 281-286.
[53] Wu, D. in Proc. *Qirt 94, Eurotherm Series 42,* D. Balageas, G. Busse and G.M. Carlomagno Eds. EETI editions, Paris, France, 1994, pp. 298-302.
[54] Wu, D.; Rantala, J.; Karpen, W.; Zenzinger, G.; Schönbach, B.; Rippel, W.; Steegmüller, R.; Diener, L.; Busse, G. in Proc. *QNDE Review of Progress in Quantitative Nondestructive Evaluation* D. O. Thompson, D.E. Chimenti (Eds.), Plenum Press, New York, 1996, *15*, pp. 511-519.
[55] Letho, A.; Jaarinen, J.; Tiusanen, T.; Jokinen, M.; Luukkala, M. *Electronic Lett.* 1981, *17*, 364-365.
[56] Bennett, C.A.Jr.; Patty, R.R. *Appl. Opt.* 1982, *21*, 49–54.
[57] Busse, G. *Appl. Phys. Lett.* 1979, *35*, 759–760.
[58] Thomas, R.L.; Pouch, J.J.; Wong, Y.H.; Favro, L.D.; Kuo, P.K.; Rosencwaig, A. *J. Appl. Phys.* 1980, *51*, 1152–1156.
[59] Grinzato, E.; Bison, P.G.; Bressan, C.; Mazzoldi, A. in Proc. *Qirt 98*, D. Balageas, G. Busse and G.M. Carlomagno Eds. EETI editions Paris, France, 1998, pp. 64-70.
[60] Meola, C.; Carlomagno, G.M.; Di Foggia, M.; Natale, O. *Applied Physics A* 2008, *91*, 685-691.
[61] Aamodt, L.C.; Maclachlan Spicer, J.W.; Murphy, J.C. *J. Appl. Phys*. 1990, *68*, 6087-6098.
[62] Tenek, L.H.; Henneke, E.G. in *Thermosense XIII Proc.SPIE* G.S. Baird Ed. 1991, vol. 1467, pp. 252-63.
[63] Rantala, J.; Wu, D.; Busse, G. *Res. in Nondestr. Eval.* 1996, *7*, 215-228.
[64] Meola, C.; Carlomagno, G.M.; Giorleo, L. *J. Nondestr. Eval.* 2004, *23*, 125-132.
[65] Meola, C.; Carlomagno, G.M.; Giorleo, L. *J. Materials Processing Technol.* 2004, *155-156*, 1132-1137.

[66] Meola,C.; Carlomagno, G.M.; Squillace, A.; Prisco, U.; Morace, R.E. *Macromolecular Symposia* 2005, *228*, 273-286.
[67] Carlomagno, G.M.; Di Maio, R.; Meola, C.; Roberti, N. *QIRT Journal* 2005, *2*, 5-24.
[68] Meola, C. *Infrared Physics and Technol.* 2007, *49*, 228-233.
[69] Meola, C. *IEEE Sensors Journal*, 2007, *7*, 1388-1389.
[70] Meola, C.; Carlomagno, G.M. *J. Adhesion Sci. Technol.* 2006, *20*, 589-632.
[71] Meola, C.; Carlomagno, G.M.; Valentino, M.; Bonavolontà, C. *2nd Int. Conference on Buckling and Postbuckling Behaviour of Composite Laminated Shell Structures*, Braunschweig (Germany) 3-5 September, 2008.
[72] Meola, C.; Carlomagno, G.M.; Lopresto, V.; Caprino, G. *3rd European Conference for Aerospace Science EUCASS 2009*, Versailles, (France) 6-9 July, 2009.
[73] Jones, R.M. *Mechanics of composite materials*, Hemisphere Publishing Corporation, New York, 1975.
[74] Hull, D.; Clyne, T.W. *An introduction to composite materials*, Cambridge University Press 1996.
[75] Richardson, M.O.W.; Wisheart, M.J. *Composites Part A*, 1996, *27*, 1123-1131.
[76] Park, R.; Jang, J. *J. Mater Sci.* 2001, *36*, 2359-2367.
[77] Shyr, T.W.; Pan, Y.H. *Compos Struct*. 2003, *62*, 193-203.
[78] Baucom, J.N.; Zikry, M.A. *Composites Part A* 2005, *36*, 658-664.
[79] Caneva, C.; De Rosa, I.M.; Sarasini, F. *Strain* 2008, *44*, 308-316.
[80] Meola, C.; Carlomagno, G.M. *Applied Physics A*, 2009, *96*, 759-762.

In: Recent Advances in Non-Destructive Inspection
Editor: Carosena Meola, pp. 125-160
ISBN 978-1-61668-550-8

Chapter 5

THERMOGRAPHY IN CULTURAL HERITAGE CONSERVATION

Ermanno Grinzato
CNR-ITC Corso Stati Uniti, 4 – 35127 Padova (Italy)

ABSTRACT

Thermal non-destructive testing techniques are more and more applied even in the field of cultural heritage conservation. The main features are the capability of inspecting large areas in a fast and reliable way, without any damage. Most of such techniques are based on the analysis of the thermal process during the quasi-steady or transient regime. IR thermography may be applied directly or under advanced processing, accordingly to particular demands. The qualitative or quantitative use of thermography allows an effective investigation of the decay, both in laboratory and in situ. Thermal Non Destructive Testing could analyze different works of art and decay forms. Results given for the following case study are illustrated: fresco detachment, structural cracks for the seismic risk evaluation, monitor of the moisture and the distribution of ambient conditions, the characterization of historic materials.

INTRODUCTION

IR thermography has been applying since '60 for several purposes in many fields. The map of surface temperature is an established method to control buildings, electrical equipments, engines, electronics, corrosion, biological activity, fire and environmental comfort. Thermography is also applied for experimental study of thermo-fluid-dynamic, for the evaluation of thermal and mechanical properties of different materials, automotive guidance etc. Others qualitative uses of IR detectors are related to surveillance, law enforcement, military and security [1].

The Non-Destructive Evaluation of materials is a very important part of our everyday life. Thermography is gaining an increasing role in this field due to its non-contact nature and the great informative content of images, but above all, because temperature and heat are so deeply involved in any physical process. The main task is finding out a testing procedure and a processing algorithm able to transform the surface temperature distribution, eventually changing in time, into the searched information. That is, to identify, locate and sometime characterise whatever is pre-defined as a defect of the part [2].

The Non-Destructive Testing and Evaluation (NDT/E) fit well with the optical nature of IR thermography and are deeply used in many industrial process. Among others, the Thermographic survey for the Cultural Heritage conservation is well documented [2, Cap.18]. Thermal Non-Destructive Evaluation (TNDE) techniques are sensitive to a wide variety of structural defects, such as voids, inclusions, delaminations, cracks, moisture accumulation, porosity and any type of flaw that determines a step or gradual change in thermophysical properties. Nevertheless, the concept of defect or even simply anomaly is not appropriate for this particular application, because very often the knowledge of the missed history of the work of art is the ultimate goal of the inspection [3]. Furthermore, the aid given by thermography to the detection of hazards, due to different risk factors as seismic activity, structural weakness, or simply the environment are very important issues for the cultural heritage conservation.

Generally, the restoration activity and the monitoring of the intervention should be based on results of combined interpretation of thermal maps with the visual images. Very often, the interpretation of results is much more challenging than for industrial issues, due to strong limitations in the test. More important, the test could be very complex due to the superposition of different forms of decay, as in case of analysing large monuments with centuries of life. In such a cases, it is needed the integration of different non-destructive methods.

The impressive improvement of the microelectronic gives two main advantages to updated applications of this method: a constant increasing thermal and geometrical resolution of new sensors and a much powerful computational power, allowing a real time processing of digital thermograms. These factors make possible the usage of complex analytical and numerical procedures to invert the thermal process. Therefore, starting from the surface temperature maps produced at frequencies even higher than TV rates it is possible to fulfil important tasks for cultural heritage. It must be stressed that thermography greatly expand its usefulness from the mere surface analysis up to the investigation of the whole body just managing the inverse thermal problem. Of course, the depth of the probed layer strongly depends on the materials thermal properties and generally is in the range of centimeters.

In transient TNDE techniques, a non-contact thermal input is applied onto a surface of the inspected body. Generally, the temperature of the surface around to the excited area is measured by infrared thermography, either during or after the thermal input application. The evolution of the surface temperature is used to assess the variations of the thermophysical properties inside the material and can provide information on the presence, dimension and depth of sub-surface discontinuities. A less practical configuration is used when the surface opposed to the heated one is analysed.

An option is the usage of natural heat sources, as the sun or artificial ones, as lamps, air guns or others, like the surface vibration.

The imposed heat flux is used as probing tool and the temperature varying on the surface is analysed both in space and time or frequency domain. A crucial point is the setting up of a suitable mathematical model for managing the thermal problem. The inverse thermal problem allows to infer information inside the body from the surface temperature or flux maps, but is more difficult than the direct one, because is generally ill posed. The effectiveness of TNDE techniques in the quantitative evaluation of structural flaws strongly depends on the availability of reliable inversion algorithms, correlating the surface response with the characteristics of defects. These inversion algorithms can be determined by analytical methods or heuristically [4,5,6,7,8].

Another combined use of optical temperature measuring and inverse thermal problem solution is shown for the measurement of thermal properties of solids and particularly of the thermal diffusivity and effusivity (thermal inertia). Thermal diffusivity and effusivity are useful not only directly in order to compute the temperature pattern, but as an informative parameter to classify material properties. For instance different metals can be ranked according to their hardness, or masonry walls depending on their moisture content.

1. Mathematical Models

Identifying a particular object in the structure can succeed only if the test is performed in the right way, that is the absorbed energy and the measuring time window must be precisely specified. For these reasons, a mathematical modelling of the thermal problem has been addressed to properly design the testing procedure. Because, there are many processing algorithms to produce reliable and easy to understand maps of the discontinuities contained in the structure, the best choice can be made on *synthetic data* provided by simulation. By using such an approach it is possible to predict capability and limits of a specific thermographic technique in a particular case. Furthermore, the inspection procedure can be optimised, taking into account actual boundary conditions. This is of particular importance when wide structures must be inspected in situ [9], as for instance an historical building.

General purpose finite elements or a dedicated finite difference code for TNDT are used simulating the temperature pattern produced on the surface by specific boundary conditions, as function of space and time.

A first example of using numerical modelling to optimize the inspection of a large fresco, painted by Romanino and located on the Malpaga castle (Bergamo, Italy) is illustrated in Figure 1a. The purpose is to scan the surface looking for detached zones. A set of four lamps heats the surface, while the thermal camera records temperature data. Finite Element Method (FEM) was adopted to simulate surface temperature using a two layer model; the first layer of 28 mm was made by plaster, the second layer (thickness 100 mm) by bricks and stone. The following data has been input for simulation: plaster: conductivity (λ) = 0.48 [$Wm^{-1}K^{-1}$], volumic mass (ρ) = 1700 [$kg\ m^{-3}$], specific heat (c_p) = 1200 [$J\ kg^{-1}K^{-1}$]; for brick: $\lambda = 0.7$, $\rho = 1920$, $c_p = 701$; for stone: $\lambda = 2$, $\rho = 2800$, $c_p = 808$. The numerical model allows to simulate the real test in order to achieve a reliable thermal signal, without any damage to the painted surface, that is reducing the heating as low as possible (in any case below 7-8 °*C*). According to simulation, the test was simulated and performed heating the surface for 150 s. Figure 1b shows a synthetic thermogram computed at maximum heating (t = 150 s). Figure 1c compares the temperature profiles by experimental data (red and more irregular line) and computed one (blue) obtained for the dotted column, marked on Figure1b, at maximum heating. It is worth noting the very good agreement (better than 0.1 K) between experimental and computed data, where the lower space-frequency content of the computed thermograms acts as a noise rejection filtering. Even the time behaviour of computed data matches perfectly with the experimental one. Figure 1d reports the temperature on the same location as Figure 1b and Figure 1c, but at a longer time (600 *s* after the start). It is evident a discrepancy between computed and simulated data on the left side of Figure1d due to the presence of a detachment that increases the surface temperature in comparison to the sound one. Actually, the defect was detected experimentally, but not included in the simulation [10].

Further details and explanations are reported in the following.

Figure 1a. The Malpaga castle (BG, Italy).

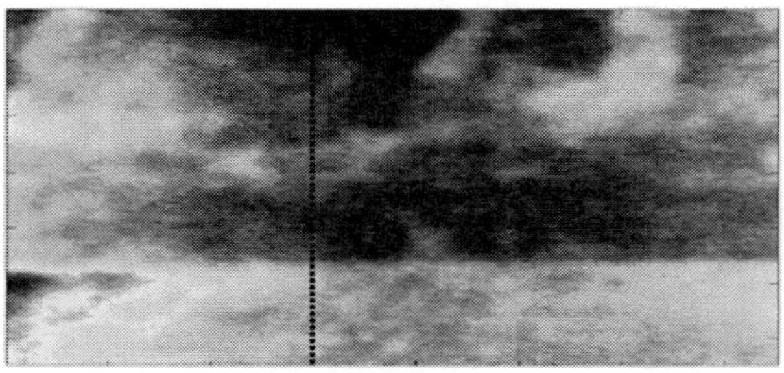

Figure 1b. A synthetic thermogram computed for the same boundary conditions and time of the maximum heating.

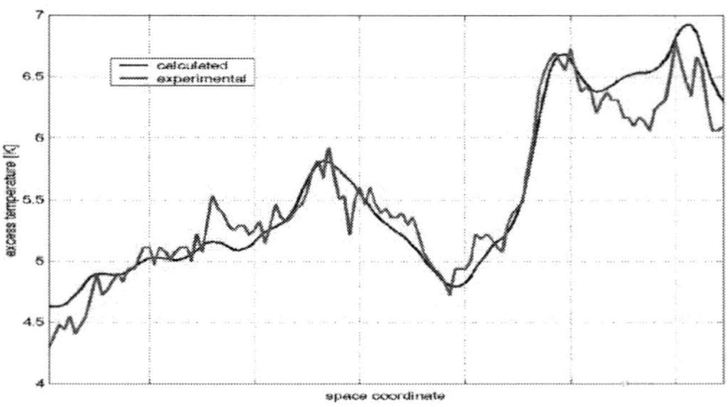

Figure 1c. Experimental and computed spatial profiles at the maximum heating (t = 150 s), corresponding to the line marked on Figure1b.

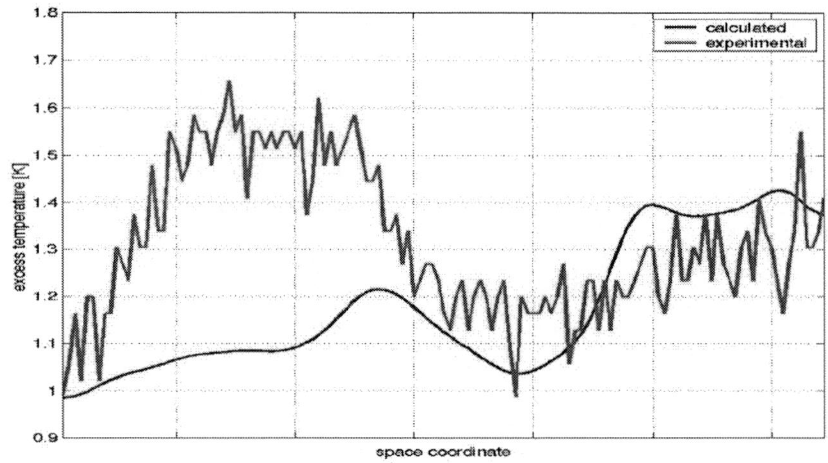

Figure 1d. Experimental and computed spatial profiles at the maximum contrast (t = 600 s), corresponding to the line marked on Figure 1b

A global simulation of a TNDE inspection procedure must take into account all the most important parameters in the heat diffusion process. These can be classified in three categories, each one reflecting a main aspect of the global problem:

- parameters related to the geometry of the simulated component;
- parameters related to thermal properties of the bulk material and defects;
- parameters related to the boundary conditions.

The component to be simulated and its constituting materials are heterogeneous and could exhibits non-isotropic properties, also varying with temperature. Several types of defect are also possible, each one showing great variability in shape, size and depth from the free surface. Moreover, the heat transfer to and from the external ambient generally happens by convection and radiation. The applied thermal stimulus can be expressed in terms of given heat flux or given temperature and can vary in space and time, the time-variation being approximated with pulsed or periodic functions.

Heat Transfer Partial Differential Equations are widely studied in Physics and Engineering. Nevertheless, only a few simple problems may be described with analytical equations in closed form [11]. Geometry of simple cases is the semi-infinite solid, the slab, the thin rod and other few cases. Moreover, the thermal parameters as thermal conductivity (λ) [$Wm^{-2}K^{-1}$], specific heat (c_p) [$J\ kg^{-1}K^{-1}$] and volumic mass, previously called density (ρ) [$kg\ m^{-3}$] are considered homogeneous, constant in time, independent from temperature and isotropic. Typically, the initial condition is a uniform temperature in the body, even if it can be neglected in case of steady and periodic regimes. Boundary conditions are natural or forced convection and radiation.

The natural boundary conditions is an effective heat source, but in order to run the test indoor or making more controlled the testing procedure it is nowadays common to impose an artificial thermal stimulation, that may be modelled as an extra heat flux term to be added to the natural one. What characterizes the different TNDE techniques and therefore the analytical model solution is the spatial and time dependence of this extra flux. Table 1 summarize a set of possible configurations.

The solution of the heat conduction equation in its general form in a heterogeneous non isotropic solid and with transient conditions can be solved as a whole by numerical methods only, which decompose the inspected system into a great number of discrete elements having simple geometry and behaviour. In particular, finite element methods or finite differences are commonly used.

The basic theory of the FEM and the other techniques for numerical analysis is well known and customized models of TNDE inspection systems can be built [21]. However, the great number of variable parameters makes it a hard task. A more cost effective way for realizing a global simulation of a TNDE procedure is provided by the commercial software products for coupled mechanical-thermal FEM analysis [22]. These offer a complete graphical environment for modeling the simulated system, managing the computing jobs and visualizing the results of the numerical analysis. Several products are currently available, which, on the basis of the specifications and the technical documentation, seem to be characterized by similar and in general, adequate capabilities for the thermal problems discussed here.

The following gives another example of using numerical modelling of the brick pattern below the plaster layer detection, that has been simulated as foreseen during a dynamic thermal test. The Figure 2a shows the surface irradiated by a set of lamps in order to inject a controlled heat flux into the wall. The temperature pattern reported in Figure 2b is a perfect footprint of the wall bonding, due to the different thermal properties of brick and mortar. Three different bricks orientations are clearly distinguished and a couple of cracks as well. Such a good result is due to a correct testing procedure, previously defined using the geometric model of Figure 2c. [23] Here, the problem is to consider different thickness of the mortar on the surface temperature map. The numerical simulation creates a set of synthetic thermograms as that shown in Figure2d, when the maximum thermal contrast has been reached. Thermograms are automatically computed by the software for any other time. Therefore, giving the temperature distribution in time and analyzing the maximun temperature difference between the brick and the mortar joints it is possible to identify when the optimum observation time happens. These values are plotted vs. time and the temperature history indicates that only at a particular time the hidden texture of the brick is detectable. The dashed line on Figure 2e indicates such a time in the case of 21 mm of plaster. Of course, this particular time strongly depends on the thickness of the plaster layer. For this purpose, different runs allow to correctly interpret complex experimental data. Figure 2f shows a temperature profile along the middle horizontal line crossing the mortar joint (see figure 2d). In such a way, we may examine the surface temperature pattern to be expected in the real test and the signal arising in critical regions i.e. in correspondence of detachments of the plaster.

The key point is the thickness of the finishing layer because the thermal process is extremely diffusive. In practice, any plaster thickness can be considered if the procedure is perfectly tailored. Unfortunately, the thickness is not generally known a priory or even. Therefore, the modelling consider this as a varying parameter. The mortar joint has been modelled as large as 10 mm, while for the plaster thickness we considered ranging from 10 to 21 mm. Mortar and plaster properties have been unified. Thermal properties chosen for the wall materials are the following: brick: $\lambda = 0.75$ [W m^{-1} K^{-1}], thermal diffusivity (α) = 5 10^{-7} [m^2 s^{-1}]; plaster and mortar: $\lambda = 0.23$, $\alpha = 2.1$ 10^{-7}.

Table 1. Analytical models for simulation of TNDE and data reduction (heat flux dependence)

spatial dependence	*time dependence*
Uniform heating Obtained heating the solid surface with lamps or air heater. Very utilized in TNDT, because it permits to analyse large areas (depending on the heating equipment). Very difficult to obtain a real uniformity in practice. For slab and semi-infinite solid produce a 1D heat diffusion along the in-depth coordinate	Impulse Realized by means of flash or switching on and off IR lamps [12]. Effective for defects detection [13]. Utilized for diffusivity measurement in slab [14]. Step (Continuous) Realized switching on/off a lamp or a forced convective heating. Used in thick slab or semi-infinite solid as a mean to determine the thermal effusivity and diffusivity. Periodic Obtained modulating periodically the power supply of a lamp, air gun heater, thermoelectric equipment or the solar cicle. Used for deeper discontinuities detection [15].
Spot Heating Obtained usually by means of a laser source heating a small area of the solid surface. It produces a 3D heat diffusion. *Moving the source it is possible to scan the surface of a solid*	Continuous It is used to detect cracks or anomalies with spatial distribution perpendicular to the surface of the solid [16]. Impulse Utilized for diffusivity measurement [17] Periodic Used for diffusivity measurement [18]
Line Heating Obtained by means of a laser source through a cylindrical lens, with a scanning mirror or a linear lamp or warm air stream. It produces a 2D heat diffusion *Moving the source it is possible to scan the solid surface. More productive of the Spot Heating technique*	Continuous Used to detect cracks perpendicular to the surface as in the Spot Heating. More productive and capable of detecting anomalies parallel to the line source or to the surface [19,20]

The mathematical predictions, shown by Figure 2d and Figure 2f have been repeated for different values of the plaster thickness. As result, a weaker signal (0.15 K) is expected for a thicker plaster layer (21 mm). Here, the importance of the three dimensional (3D) analysis is also evident, indicating the boundary of the area where a 1D heat flux is approximated. The 1D region is limited to the flat bottom in the temperature profile of Figure2f. This is important, because only within 1D area usual inversion algorithms are correct.

Figure 2a. the tested surface (the vault) with the lamps used for the heating.

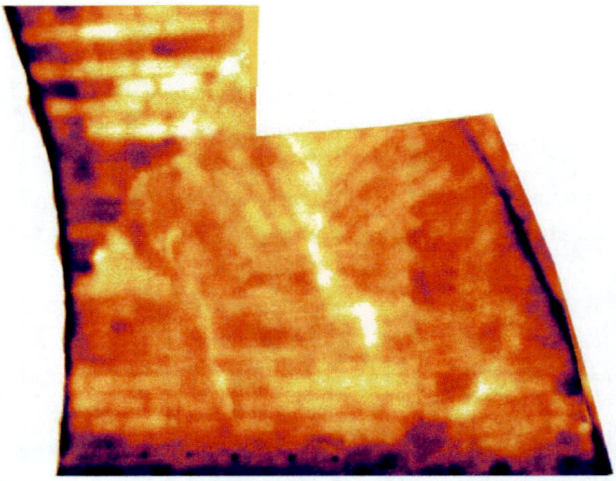

Figure 2b. thermogram of the surface giving the wall bonding and a set of cracks below the plaster.

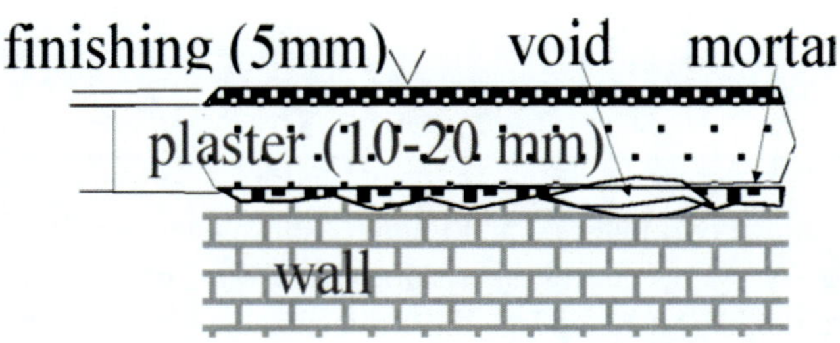

Figure 2c. model of the wall used for the numerical simulation of the test for the wall bonding detection under the plaster.

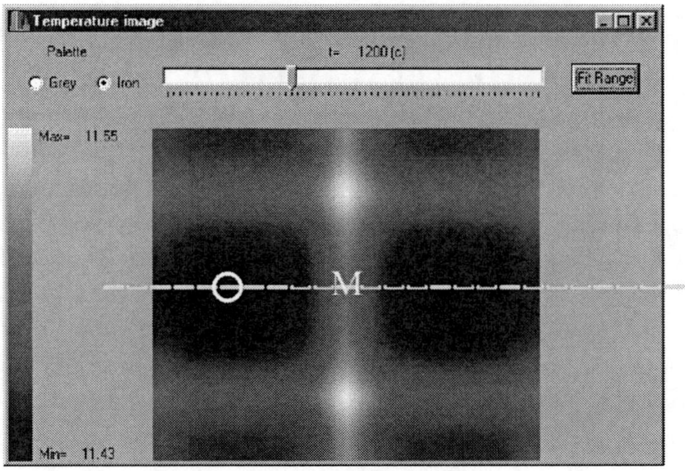

Figure 2d. synthetic thermogram with brick joints, below 21 mm of plaster computed at the time of the maximum thermal contrast.

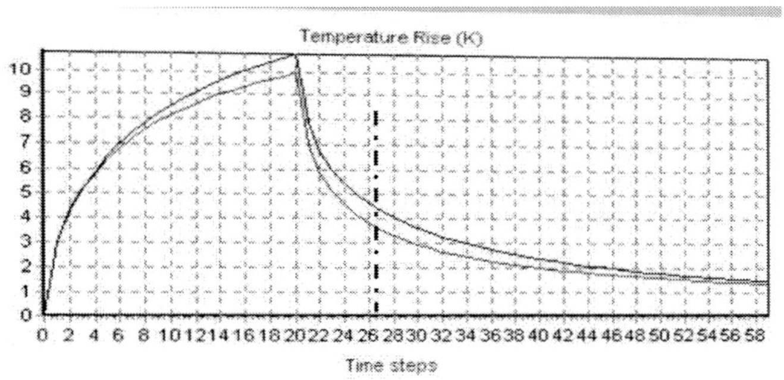

Figure 2e. Time profile of surface temperature over brick (O) and joint (M) as foreseen by the computed thermograms' sequence (plaster layer 21 mm) the dashed line is the time of maximum contrast).

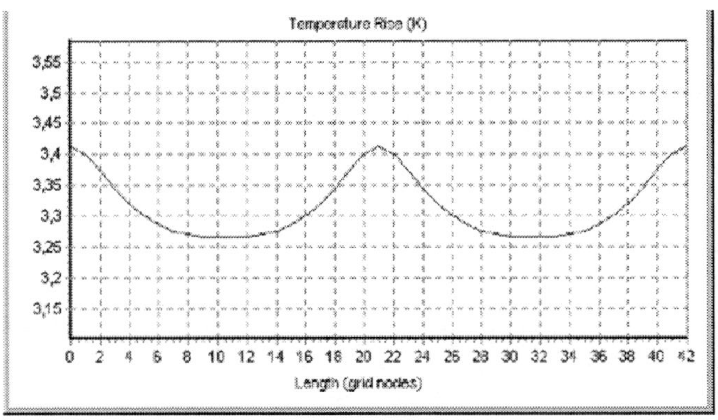

Figure 2f. Spatial profile of surface temperature as foreseen at the optimum time, across dashed line of Figure2d, the synthetic thermogram of brick joints.

2. From the Surface Temperature to the Control of the Cultural Heritage

IR thermography has been applied for more than 40 years to the buildings monitoring in qualitative or quantitative way [24,25]. Generally speaking, there are many possibilities to transform the temperature mapping of the surface into the investigated parameter. Up to now, IR Thermography has been applied successfully by the authors for the identification of cracks and detachment of frescoes, the identification of hidden structure or the wall bonding knowledge, the moisture mapping of historical buildings, the testing of wooden icons [26] and the analysis of archaeological pottery. Finally, the study on the internal microclimate has been also accessed [27]. In theory, any step or even progressive changes in thermophysical properties, related to the presence of materials discontinuity or ageing may be evaluated when an appropriate thermal flux is imposed to the object. In practice, the signal due to the discontinuity is very low (in the order of tenth of Kelvin). To achieve such a high level of detectivity a special set of hardware and software tools have been developed and implemented. It is worth saying, that Thermal Non-Destructive Testing or Evaluation are highly sensitive to a wide variety of noise sources, as most of inverse problems. The electronic noise is not the only one, but also the variation of the surface in emissivity, shape, colours, composition etc. or the evenness of the heating generate serious disturbs.

In any case, a heat flux must cross the envelope in order to produce a temperature distribution on the surface to be detected by the thermal camera. If the heat flux is generated by natural boundary conditions, the inspection is called passive. Often, the natural heat fluxes and temperatures vary slowly, because of the high thermal inertia of building materials. In the case of quasi-stationary thermal process a single thermogram is processed or the average of a few tehrmograms, just to reduce the random noise. At this stage, a common request is the geometric composition of different thermograms, with the needed correction for the perspective distortion, in order to have a total view of the building. This process is not trivial, because the quantitative temperature information must be preserved during the images processing [28]. A complete scanning of wide surfaces using the natural heat source is easier from the point of view of the needed equipment, but attention must be paid to geographical aspects and topology of the building.

On the other hand, transient techniques can be applied using an artificial heat source. This is generally the case, when inspecting indoor. Guidelines for the inspection procedure using a transient thermal test involve the following steps:

- heating uniformly the surface using radiative or convective heat transfer, up to a few Kelvin above room temperature;
- storing the sequence of thermograms by using a computerised thermographic station;
- processing the whole bunch of stored data with a algorithm for transient thermal NDT.

Thermal analysis is normally performed by processing data off-line. Either by qualitative or quantitative analysis is possible detecting large debondings and cracks, not visible on the surface. Generally, the temperature history of any point is compared with a reference one, chosen on a sound zone. Therefore, if a reference area is needed, a preliminary visual analysis and tests using alternative methods are useful. A simplified technique is detecting of discontinuities analyzing a single thermogram, taken at the optimum observation time. A quantitative evaluation requires processing of the whole thermographic sequence. In such a way, the discontinuities characterisation is normally fairly accurate, even if the algorithm is time consuming. Furthermore, the quantitative processing algorithms are more complex and under development.

A thermographic sequence always contains:

- the initial image taken at ambient temperature;
- the image of maximum temperature taken at the end of heating;
- a set of images taken longer than the optimum observation time for the deepest discontinuity of interest, recorded at the proper time step.

Many analysis techniques are available to extract detailed information from the thermograms. Among different algorithms available *Dynamic Thermal Tomography* (DTT) [29,30], *Pulse Phase Thermography* (PPT) [31] and *Principal Component Thermography* (PCT) [32] have been preferred. Not only the data reduction among them is different, but also the common way to inject the heat flux onto the surface. Actually, the simplest form to inspect a body is to analyze the temperature field given by a heat pulse. Unfortunately, the low thermal conductivity of porous and fibrous materials requires a relative long heating in order to probe a depth of a few centimetres. Hence, it is not trivial the way to manage data when a heat flux lasting for a few minutes is used instead of a sharp pulse [33]. So that, DTT is tailored mainly for the finite pulse heating; PPT is more used for pulse or modulated heating (even in non periodic regime) and PCT for irregular heating.

The first procedure, called DTT analyses in time domain the normalised thermal contrast. The contrast is defined as the temperature difference between the point and a reference point, normalised by the reference temperature. Several variants of the contrast exist, for instance the maximum temperature reached by any point is used for the normalisation. In such a way, the influence of different heat absorption on the surface is reduced. The analysis of the whole set of data gives two synthetic images, where the maximum of the contrast is mapped, concurrently with the time when it occurs. This procedure characterises the defect, using proper calibration functions. According to this, a window in time evolution corresponds to a layer in depth. This allows separating the searched defect from the whole structure. The main advantage of DTT is the relatively simple evaluation of the detachment depth [34].

The other technique, knows as PPT is based on the discrete Fourier transform applied to the time-domain of the thermograms sequence. The final result is represented by two sets of magnitude and phase images, each one being associated to a frequency in the range [0 - Δt^{-1}], where Δt is the sampling interval. Among the advantages of this technique we mention the less sensitivity of the phase map to the optical and geometrical characteristics of the surface and the fact that it is not required indicating a sound region, therefore, it can be fully automatic. Furthermore, the probing depth of phase is higher than amplitude based algorithms [35].

PCT is a variant of Principal Component Analysis [36] that is an algorithm that acts on data exploiting their statistical characteristics. The covariance matrix of the data is computed, representing the variation of the temperature history profiles for each pixel, with respect to the average temperature profile. Therefore, a new data set of the temperature increasing is produced, by distributing the informative content in decreasing order with respect to the thermal variation occurring during the test. The vectors (eigenvector) that maximize the variation in the data are assumed as new axis of a orthonormal system of coordinates. Such a rotation of the reference system allows the new system (principal components) to be directed along the maximum variation direction. The original data set is finally projected on the new axes, obtaining maps with enhanced visibility of the defects. PCA is used also as a tool for strongly compress data, because it is possible to use only a few components to represent the chosen amount of information. In the cases of the tests here considered, typically four components represent 99.5% of information. Results on samples showed that the predominant temporal variations of the entire data set are usually contained in the second and third eigenvector. The PCA processing, even though not quantitative, strongly reduces false alarms given by the very effective, but uneven heating provided by the hot air blower. Another use of PCT is for denoising data, at first applying directly the algorithm, followed by an inverse step resulting in a termographic sequence reconstruction.

A strong interest has been devoted to the monitoring of heritage buildings covering layers and particularly the fresco adhesion status. The detachments or lack of adhesion between the finishing surface and the substrate are very common. The parget layers may embody an air gap inside or between the wall and the plaster. They are due to bad setting of the plaster or the crystallization cycles of soluble salts. The detection of the defect is given by this discontinuities acting as thermal barrier. Also in case of strong tensions giving structural fissures the plaster is normally detached from the background, even if the surface is not jet cracked. Detachments between frescoes and supporting walls have been analysed both in laboratory and *in situ* [2,3,10,34,37].

A common way for the artificial heating is by using a set of IR lamps delivering about 3000-8000 [W m^{-2}] for 100-300 [s] or with a moving linear heat source [38]. Thermal tomograms manifest well the detachment between fresco plaster of 1-30 [mm] thickness, in case of uniform plaster surfaces. In real case, a heating pattern is always superimposed to the thermal signal. However, the main problem is the fresco itself producing uneven energy absorption due to varying colours [37]. The uneven heating creates fake signals, that obscure visibility of subsurface discontinuities. The contrast normalisation with respect to maximum temperature is typically applied to reduce uneven heating phenomena. This is helpful, but does not solve completely the problem, because it takes into account the propagation of heat flux in only one dimension (1D).

A raw thermogram, extracted from a sequence of 600 images is presented in Figure 3c, where the higher intensity of the lighting near the corners is well visible. The uneven heating is smoothing in time due to 3D heat diffusion that is not taken into account by conventional normalisation at the end of heating. But lateral heat flux is a source of false alarms.

A way to get rid of such a problem is solving the three-dimensional (3D) transient heat conduction problem allowing synthesize artificial image sequences that can be used to normalise source sequences, thus eliminating lateral heat diffusion phenomena. The underlying idea is that normalisation procedure can be thoroughly improved using instead of a single experimental image a whole simulated sequence that corresponds to sample temperature in the absence of subsurface defects. Simply speaking, the subtraction technique deals with the normalised difference between experimental and simulated data, while the ratio technique is based on taking their ratio. The former is preferable when the thermal parameters and geometry are well known, allowing an absolute correction. The latter gives a reduction of the 3D heat diffusion effect [10].

Figure 1 previously presented exemplifies of application of transient TNDT to cultural heritage experienced on an Italian castle. The fresco section is shown on Figure 3 together with the heating device. Figure 3 also summarize the PPT algorithm working in the frequency domain, which process a set of raw thermograms and detects a detached area in the phase image as a darker spot (high phase lag). It is interesting to note that another detached area (deeper and probably overlapping each other) is indicated by a lighter colour on the left side of the shallower detachment, because of the different depth. Finally, comparing the detachment detection given by PPT (see bottom-right corner of Figure 3) with the thermogram (Figure 1b) clearly demonstrates the great advantages in terms or reliability given by the sequence processing. It is worth noting, the phase image is less affected by optical noise, e.g. the emissivity varying, than direct observation or other amplitude related processing [35].

Figure 4 reports of using the PCT algorithm to test a fresco inside the San Francesco church located at Carapelle, Italy (Figure 4a). Here, the target is to help the evaluation of the seismic risk, by means of a precise identification and classification of the masonry [39]. The heating by means of warm air is shown in Figure 4b. A sequence of 192 thermograms has been taken at 5 s time interval, only during the cooling phase. The 2nd component of PCT is shown in Figure 4c. The processing allows finding out: the shape, type and size of elements assemblage.

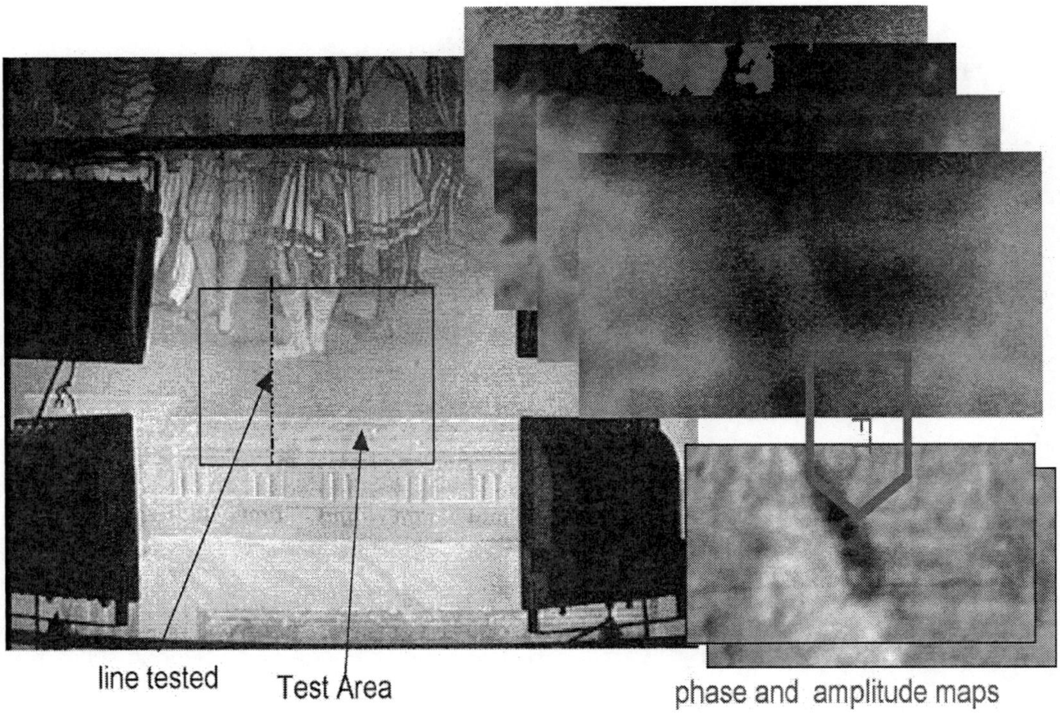

Figure 3. test of a fresco using PPT: from a sequence of thermograms (top-right), are computed the map of maximum phase and amplitude contrast, through a frequency domain processing (bottom-right), where detachments of the fresco are clearly seen.

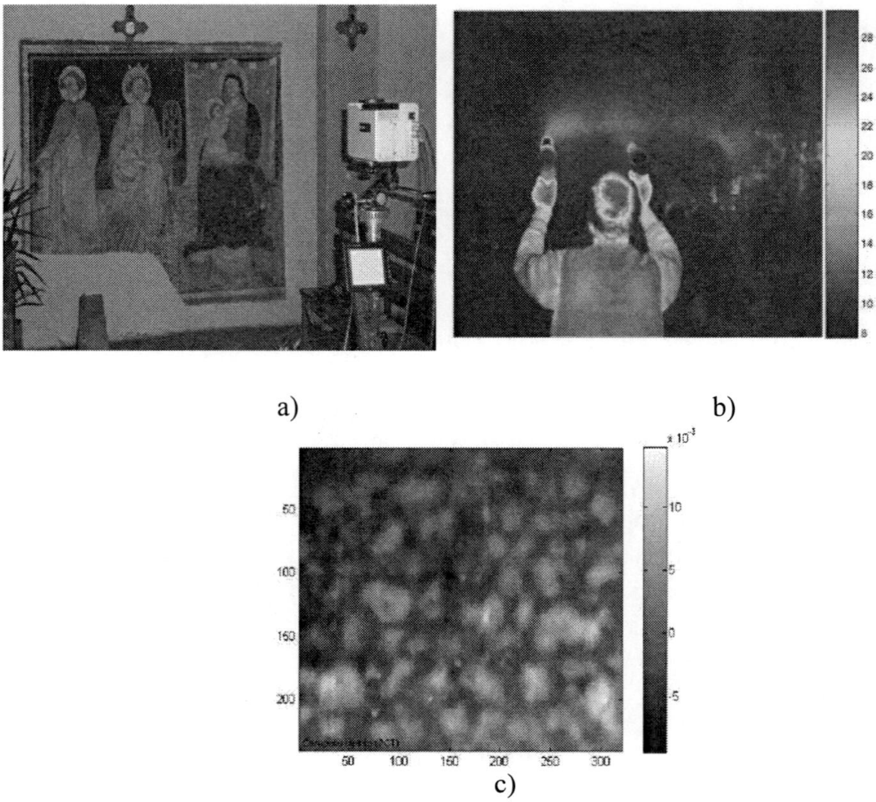

Figure 4. TNDT by PCT: picture (a); heating phase (b); resulting wall bonding (c).

Another important test for the evaluation of the seismic risk is the detection of hidden details of the building and structural cracks. Figure 5 shows the thermographic survey recovered on S. Stefano Church, Sessanio, Italy, at April 2008 [39]. In Figure 5b there is the picture of the façade. The thermal image in Figure 5b is a composition of 16 different thermograms. A dedicated algorithm called High Dynamic Range Thermography [40] has been used to enhance the structural weakness points. Here, a network of cracks on the surface of the façade clearly indicates the excessive load of the tie applied to the wall. In fact, the cracks starting points are located either to the corner of the window or to the "nails" closing the tie. Figure 5 also proves existence of two massive wall leaves for each side, not linked each other. It is interesting to point out how useful is the comparison with the picture. In addition, it is worth saying that cracks are not visible in the photograph, because very thin or not affecting the plaster layer. The reason for the presence of the footprint of such defects on the thermal image (characterized with a much lower spatial resolution) is probably due to the local delamination of the plaster from the masonry, due to the mechanical strength applied. A deeper investigation on this topic is ongoing. The final results of thermal scanning and the picture of damages suffered by the church, after the earthquake of April 6th 2009 (Figure5c), match in an impressive way. All detected defects, including a closed window below the present one, the cracks pattern, etc. have been detected correctly.

So far, heat transfer modelling, using different heat capacity of materials has been used to find out the hidden bonding below the plaster layer or closed openings. Strictly speaking, these information are not related to NDT, but are useful for the knowledge of the building and the restoration activity. Figure 6a shows a wall partially irradiated by the sun. Figure 6b has been obtained rearranging raw thermograms in time and mapping the angular coefficient coming from the linear fitting of each point temperature increasing, vs. the square root of time. Here, both characteristic river stone shapes used to built up the wall and a closed windows are visible [2 cap.18]. This inferred the age of this part, as belonging to the original castle settlement.

a) b)

c)

Figure 5. (a) the façade the S. Stefano Church, Sessanio, Italy at April 2008; (b) thermal image composition detecting of crack pattern on and two massive wall leaves for each side at April 2008; (c) the picture of damages suffered by the church, after the earthquake of April 6th 2009

Figure 6a. Visual images of the South-West walls at second floor under solar irradiation.

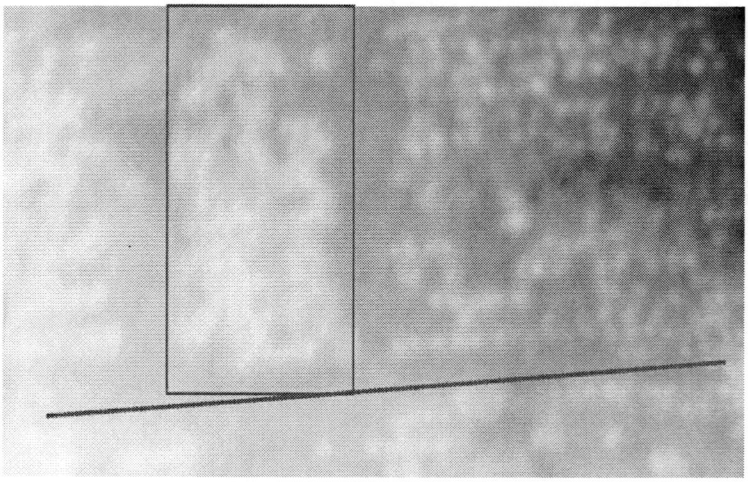

Figure 6b. Thermal analysis of the region marked on Figure6a based on thermal inertia local mapping of the South-West walls indicating the hidden bonding and a closed window.

Another issue is the Non-destructive moisture detection, because moisture is the main reason for building damages. Current building moisture surveys performed using equipment such as moisture meters are slow and prohibit surveying hard to reach locations. An infrared thermal imaging camera helps to monitor moisture by imaging the different temperatures of wet versus dry building materials, making it a useful tool for a building moisture survey. Areas surveyed by the IR camera can be located before standard moisture detection equipment can. Sometime, a moisture survey of water damage can be performed with the infrared camera simply by walking through the affected building. Nondestructive water intrusion testing can be performed using the IR camera in modern buildings, but could fails when high relative humidity and low temperature decrease the evaporation process. More important, the relationship between moisture and temperature is much more complex as expected and a much deeper knowledge is needed [41].

There are at least three important physical phenomena correlating temperature to the water content of a porous material in a significative way. The first is the evaporative cooling, reducing the surface temperature accordingly to the flux of water vapour released and the related latent heat. The second is the thermal inertia of a moist porous material, that is much higher than the dry one [42]. The third is the characteristic optical absorption spectrum of water.

The next example refers to the moisture detection on the roof of a heritage building in order to check the drainage capability [43]. Also in this case, the higher thermal inertia of the water than the dry building material is examined during day and nigh time. Furthermore, the cooling effect acts the most important role in determining the surface temperature.

The Sengul Hammam (Ankara, Turkey) is a typical Ottoman double historical baths (hammams) consisting of two separate parts for men and women. It was constructed with stone masonry walls with brick transitions and brick upper structure. The hammam includes the tepidarium and caldarium sections, hot water storage and firewood storage rooms. The roofs above, including the dome surfaces, were repaired with an addition of 8 cm thick mesh-reinforced concrete layer. The roofs of the frigidarium sections at the North and South are timber pitched roofs covered with fired-clay roof tiles. The immediate periphery of the structure was totally covered with asphalt pavement. Unfortunately, a severe degradation of the building followed such a bad restoration. The restoration intervention modified the original roof drainage system in terms of their performance, capacity and adequacy, that have been kept properly functioning for a long periods of time.

The roof areas were examined by using some non-destructive methods. The maps of visual decay forms observed on the building facades were produced in Figure 7. The analysis of the decay in the structure and their distribution is indicative of the probable sources of these problems. Following this preliminary on-site visual survey, a planimetric survey was conducted to record the topographical features of the roof. Readings were then converted into a detailed map indicating overall slopes and local falls in reference to roof drainage components (waterspouts and downspouts). This made it possible to locate problem areas with a potential risk of ponding on the map. Dimensions of all spout outlets, eaves gutter and downspouts were also taken to facilitate capacity calculations. A thermal monitoring to locate damp zones scanned the areas under study, together with the lower and upper parts of the exterior walls. Special attention was given to the lower parts of the walls at points, where roof drainage components existed overhead. Damp zones were compared with the risky areas for ponding. Drainage calculations were made to assess the discharge capacity of the roof and their adequacy [44, 45]. The maps of decay forms and infrared images showed the problem areas of continuous rainwater penetration and their location in the structure. The severe material loss on masonry surfaces together with staining and salt deposits were found to correspond with the damp zones visible in figure 8 as colder. Those surfaces were exposed to rainwater, overflowing from the eaves gutter, downpipes and waterspouts. Hence, the lower parts of the walls on the axis of downpipes and waterspouts were found to be damp and cold in the infrared images of Figure 9a. The height of cold areas was detected to extend towards the discharge components above (see Figure 9b). The periodical wetting and drying at those damp zones must have caused such severe deterioration on wall surfaces. The recent repairs with cement mortar were also introduced salt problems accelerating the deterioration mechanisms at the presence of dampness.

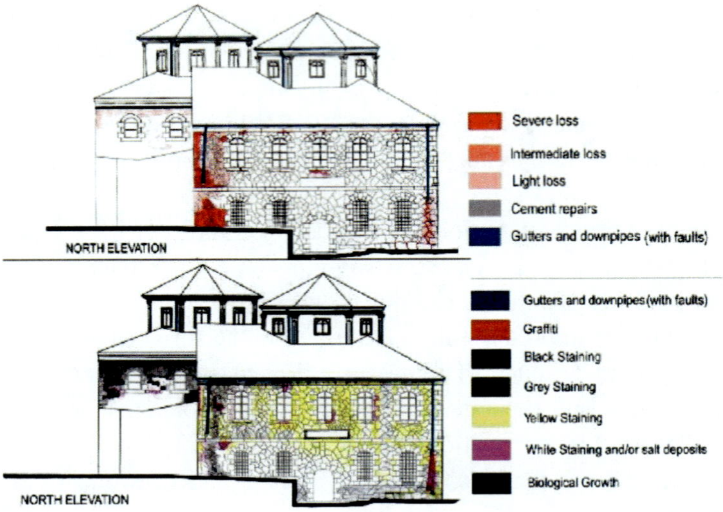

Figure 7. The maps showing the distribution of material loss (at top) and the distribution of discoloration and deposits on surfaces of the north façade (at bottom). The severe loss and detachment on surfaces was observed together with white staining and salt deposits, especially at the lower parts of the rainwater pipes at both sides of the façade and stone courses at the top, underneath the eaves gutter.

Figure 8. The infrared image of the selected region at the North façade shows the severe material loss as warmer areas and the detachments, which were not observed visually as colder areas, both suffering from the roof drainage faults.

Figure 9a. The detached and moist areas close to the rainwater downpipes were detected as colder areas.

Figure 9b. The lower parts of the wall corresponding to the axes of the downpipe and the waterspout were found to be damp.

The overloaded spouts could also be detected in the IR images of Figure 10. Actually, almost all metal components were found after this survey to be rusted and lost their functions.

The present conditions of the roof were determined to be unsatisfactory due to unconscious interventions and poor maintenance. In particular, the repair with mesh reinforced concrete layer added on the roof was unacceptable, since it was introducing salt problems to the structure and triggering the deterioration mechanisms in presence of dampness. This heavy concrete layer also increased the dead load of the structure and weakened it against the lateral stresses. Most faults observed on the roof surfaces were inevitably the result of this concrete layer.

Figure 10. Colder areas in the IR image of the selected region indicates the damp zones at the upper parts of the wall due to the overflowing from the roof eaves level between the spouts WS6 and WS7. Severe material loss together with salt deposits and biological growth overlap with the cold areas

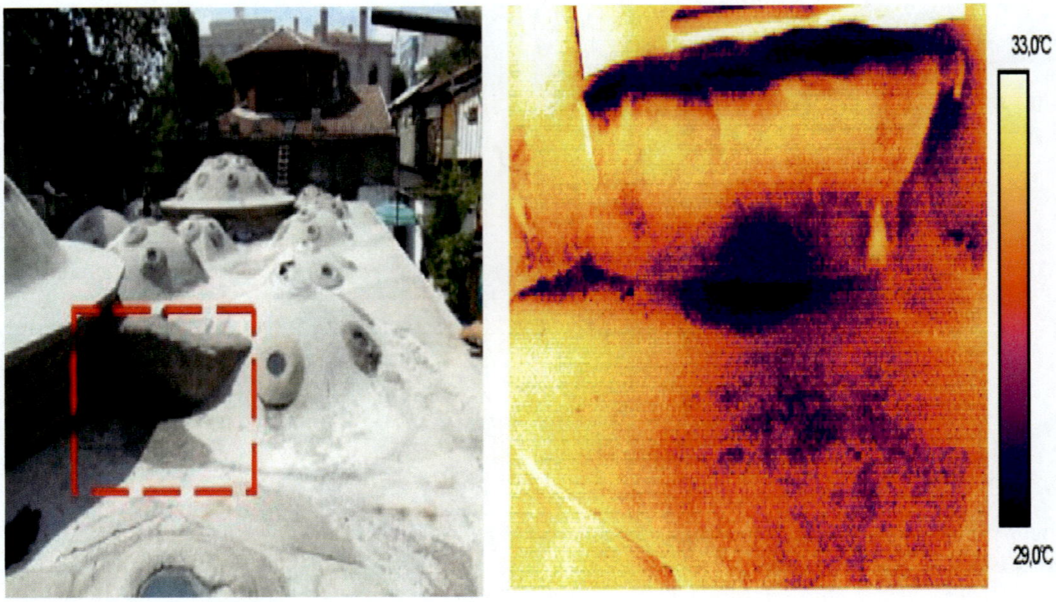

Figure 11.The IR image of the parapet cutting the water flow and causing soil/dirt accumulation in front of its outlet. The gradual decrease of surface temperatures towards the outlet exhibited the potential for absorbing and retaining rainwater.

The indoor microclimate is very often the major hazard issue for the conservation of cultural heritage, including wall and wood paintings. In many cases the heating devices are requested inside the monuments, but it is difficult to forecast the long-term effect on precious work of arts. Generally, a year round lasting measurement of the air temperature and relative humidity, coupled to visual inspection and photographic recording is the way adopted for safeguarding the monument. Sometime, additional methods are applied as for example IR thermography, but not on regular basis, due to lack of precise procedure and standards. Furthermore, the cost is a limiting factor to only a few case studies.

Existing instruments to monitor environmental conditions are not appropriate to map local values such as temperature, light, humidity, air velocity inside buildings. For instance it is totally unpractical to move a measurement head comprising a temperature sensor, anemometer, hygrometer every 10 cm in a given room! Moreover, sometimes variations of interest are below the range of common instrumentation. For instance, low limit for ordinary anemometers is about 0.5 m s^{-1} while natural convection occurring in buildings could be even smaller than 0.1 m s^{-1}. Tight building envelopes are another rising problem since, in case of poor air exchange, risk of higher level of humidity might conduct to health risks related to developed fungus.

The study of the indoor thermal and humidity conditions is a problem addressed by many authors, since a long time ago [2, 24, 46]. The distribution of the surface temperature is of course very important, but the correct approach is to measure the boundary air conditions, at the same time. It is a matter of fact that the main limiting factor is the difficulty to analyze the environmental conditions using a suitable time and space scale. In fact, for a detailed study the density of measurement required is much more extensive than the ones practically available using ordinary techniques. Furthermore, the high accuracy of measurements is hardly matched at a reasonable cost. Really, the temperature of the surface could be measured with contact probes, but the contact thermal resistance strongly reduces the accuracy. Therefore, IR thermography is a good way to fulfil such a task. It is worth saying the low gradients to be measured, requires special tools in order to achieve an accuracy of the order of 0.1 K. Finally, the normal indoor conditions are affected by an operator, during the measure itself. For these reasons a totally automatic measuring system is the best solution to the problem [47].

The following case study illustrates a new approach to solve such a problem (patent pending) [48]. The main issue addressed is the automatic measurement of different ambient parameters, taking advantage as much as possible of thermal images. Basically, the new deal is to measure the surface and air thermodynamic status starting from temperature measurements on a special set of targets, distributed on the inspected volume. Finally, the local conditions are recovered by means of an automatic processing, based on a robust mathematical model.

The final goal is to make such a crucial phase for the monument life fruitful, cost effective and reliable. It is based on two steps: at first a short run monitoring using a new Quantitative IR Thermography (QIRT) technique; then the modelling of different occupancy policies and Heating, Ventilating and Air Conditioning (HVAC) solutions. An additional step could be added for verifying of the plant efficiency and tuning it.

For all these reasons, the scanning of the wall surfaces, floor and ceiling by the IR camera is the correct approach in order to have a high-resolution detailed map of the physical phenomena distribution. The automatic scanning device detects the targets in the right order, even if they are not in a regular pattern, in order to follow the geometry of a complex building [47]. After that, the camera focus a partial fields of view and record the images. The needed values are extracted and all fields are added up to a total view of the surface. Within any elementary field of View (EFOV), the thermodynamic quantities are evaluated. Finally, a resampling process gives a very detailed spatial distribution of the air and surface thermal hygrometric conditions. Of course, the arrangement of different thermograms composing a mosaic of the entire wall surface underlies the hypothesis of quasi stationary conditions. Therefore, the faster the scanning, the better is.

As case study as been selected the church of S. Biagio (Godega di S. Urbano, Italy) with a precious XIII century fresco. The small building is made as an aisless halls, with a projecting apse at the chancel end. The monitoring of the church has been performed on December 2008. The results of the thermographyc survey has been overlapped to the 3D model of the building in order to achieve a clear diagnosis of the apse fresco decay. Figure 12 shows the temperature map of the North wall, where a cold area has been detected on the upper part, due to the top limit of an attached shelter. A cold air leakage through a wooden door has been also detected at the nave-apse intersection. Analyzing the air temperature (10 cm far from the wall) shown in Figure 13a it is clear the role of the window of the South side, which push air during the day towards the East side. The air movement given by the new thermographic technique is reported in Figure 13b. It is clear that the warm air from the people during ceremonies may easily condensates on the cold surface and this contrast the natural laminar flow. The resulting effect is a very poor ventilation (velocity$<$ 0.2 m/s) of the East wall shown in Figure 14b (beyond the altar) and a decay of the fresco. Actually, the more deteriorated fresco is marked as dotted red area in Figure12 corresponds to the coldest surface.

The second step evaluates effects of the number of people inside the church in critical winter conditions. Furthermore, a radiant heating floor hidden on the benches platforms and eventually on the altar platform has been simulated. Figure 14a shows the air temperature at 9 cm height from the floor, as simulated in case of only nave heated. Figure 14b shows the air movement on the church cross section set in the middle of the bench platform. The warm air is pushed towards the apse and stopped abruptly by the East wall in a cascade effect that makes the surface condensation possible. Adding the altar platform heating, this effect is significantly reduced.

In conclusion, the new approach allows a fast diagnosis of the cause of work of art deterioration, when the indoor environment may become armful. More important, different remedies are studied and validated by the double evaluation: experimental and numerical achieved by QIRT and Computational Fluid Dynamic (CFD), especially the air temperature and speed.

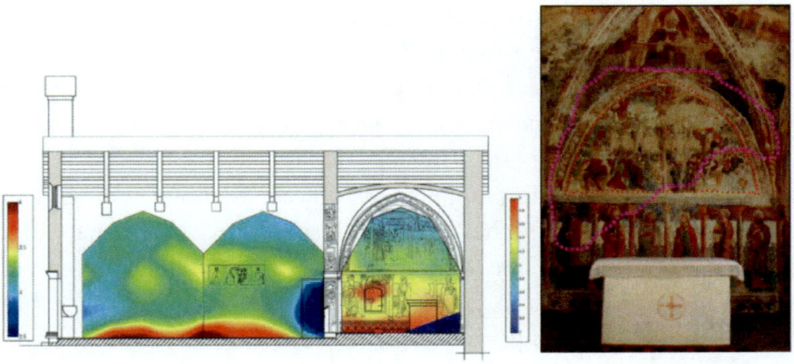

Figure 12. Surface temperature of the North wall (S. Biagio church) registered on the building model and the decayed fresco of the East wall (right); the red marked area corresponds to the coldest surface of the wall, where condensation happens.

Figure 13a. Air temperature of the apse at 10 cm from surface; thermal effect of a window on the South wall and cold top zone on North wall.

Figure 13b. Air speed measured close to noon on the apse 10 cm from the wall; the closed window on the South side thermal influence.

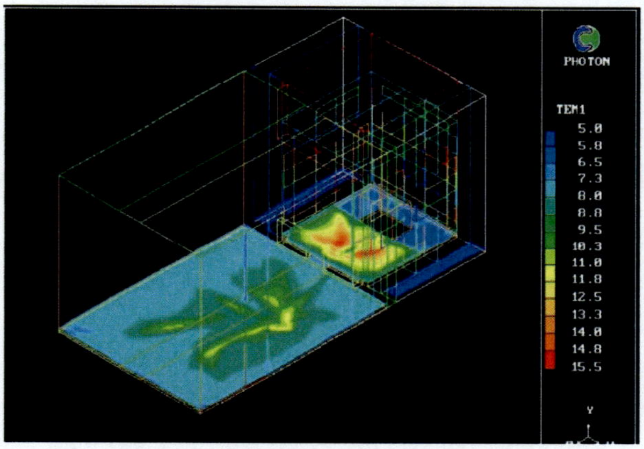

Figure 14a. Air temperature computed 9 cm from the floor with a heating radiant platform on

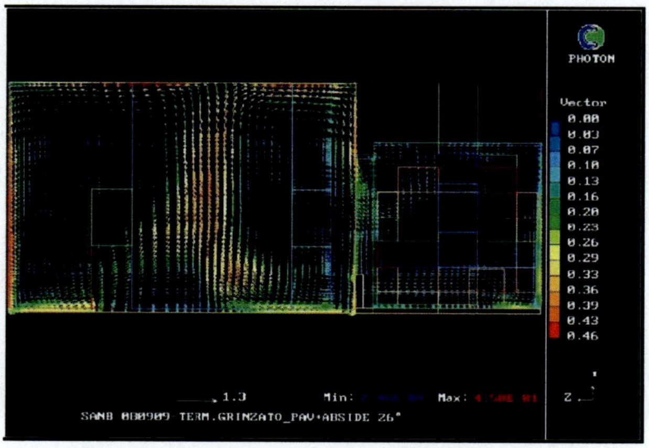

Figure 14b. Air speed computed on a vertical plane over the heating plant on the church floor.

3. Characterisation of Cultural Heritage Materials

Recovering quantitative information about the internal structure of an object from the surface temperature evolution requires the knowledge of some material properties. The usage of mathematical modelling requires a precise knowledge of the thermal properties of materials, as well. However, in practice there is a large variability, due to previous repair and restoration. From the other side, cultural heritage objects are generally very old, therefore precious and delicate. It is a mater of fact that only NDT are suitable for such a measurement. Thermography can well suit these requirements and some testing procedure are well established for measurements, both in laboratory and even in situ [49].

In dynamic thermal tests one of the most important parameters is the thermal diffusivity α, expressed in m^2s^{-1}, that is a measure of the rate at which the heat is diffusing through the material. The key role of thermal diffusivity in solving the direct and inverse dynamic thermal problem makes its knowledge necessary in the most of cases. Unfortunately, the literature data do not help enough especially for ancient works of art due to the large range of reported values. The best is the direct measuring of the most important parameters as thermal diffusivity α and effusivity (e).

3.1. Evolution of Diffusivity Measurement by IR Thermography: The Flash Method

At present the standard method for measuring thermal diffusivity [m^2s^{-1}] is the *flash method* proposed by Parker in 1961 [14] and become an ASNT standard in 1992. It is based on the analytical solution of Fourier equation with adiabatic boundary conditions for an infinite slab subjected to a thermal Dirac pulse. The test is performed according to the transmission scheme: the sample is located between the heater and the sensor. An energy pulse is deposited on a sample surface (called *front surface*) and the temperature evolution is measured on the other surface (*rear surface*). Thermal diffusivity is linked to the sample thickness L and to the half-maximum rise time $\tau_{1/2}$ through the following simple equation:

$$\alpha = \frac{0.139 \cdot L^2}{\tau_{1/2}} \tag{1}$$

The test requires a complex apparatus to ensure that the theoretical assumptions are met. To bring this method into field applications of IR thermography, the most restrictive requirements are the need to stimulate the sample with a short duration pulse and the necessity to carry out the test in adiabatic conditions. Indeed, testing a low diffusivity object generally requires a large amount of energy and therefore a longer duration pulse. Moreover, if the test is carried out in a non-controlled environment, the surface heat exchange must be taken into account. In such conditions eq.(1) does not provide correct results and a more sophisticated model is needed.

The analytical solution for the problem described above is available in many textbooks on heat transfers [11]. We resume here the two Fourier equation used for the heated side and the rear side of of an infinite slab of thickness L subjected to a heating square pulse of duration τ_h (s) and power density Q [Wm^{-2}], corresponding to the two used experimental layouts:

Heated side:

$$T(t) = A \cdot \sum_{n=1}^{\infty} \frac{2 \cdot \sin(\mu_n)\cos(\mu_n)}{\mu_n + \sin(\mu_n)\cos(\mu_n)} \exp\left(-\mu_n^2 \frac{\alpha t}{L^2}\right)\left(\exp\left(\mu_n^2 \frac{\alpha t_h}{L^2}\right) - 1\right) \qquad (2)$$

Rear side:

$$T(t) = A \cdot \sum_{n=1}^{\infty} \frac{2 \cdot \sin(\mu_n)}{\mu_n + \sin(\mu_n)\cos(\mu_n)} \exp\left(-\mu_n^2 \frac{\alpha t}{L^2}\right)\left(\exp\left(\mu_n^2 \frac{\alpha t_h}{L^2}\right) - 1\right) \qquad (3)$$

with T temperature increase, t time, $A=Q/h$, μ_n the positive roots of the equation $\mu_n \cdot tan(\mu_n) = Bi$.

The diffusivity value is obtained as a result of an optimisation procedure (for instance least squares fitting) based on Eq. (3). In addition to the estimate of α, the procedure provides estimates of (Q/h) and the Biot number ($Bi=hL/\lambda$), h is the surface heat exchange coefficient [$Wm^{-2}K^{-1}$].

Further improvements concern the actual shape of the heating pulse that sometimes differs very much from the supposed square shape. Indeed, the pulse released by a lamp or flash tube involve rise and fall times that sometimes cannot be neglected. The proposed solutions to this problem are based on convolution in time domain or deconvolution in frequency domain [50].

This method has been successfully applied to a wide variety of materials: composites, metals, plastics, building materials, etc. The application of the flash method to the diffusivity measurement of historic bricks, plaster samples or archaeological pottery requires further precautions. The most important is the non-parallelism of surfaces, which are so fragile not supporting any mechanical work. For this reason, using thermography we can virtually process many parallel tests at the same time, getting rid of such a problem trough statistical algorithms.

Two sets of measurements have been carried out on two samples coming from a historical arsenal buildings located in Venice (Italy): a piece of brick 5.5 mm thick and a piece of plaster with a thickness of 5.6 mm. The samples, with just roughly parallel faces were heated by means of a pulse of duration ranging from 1.5 s to 2.5 *s* and the temperature evolution on the rear surface was observed through an infrared camera sensitive in the 8 μm – 13 μm wavelength band. The sampling interval was set to 0.5 s and the observation interval to 60 s. Hence, for each measurement a sequence of 120 thermograms was acquired.

Figure15a and Figure15b show both a raw thermogram and the average temperature history measured on the rear side of the sample during one of such tests. The temperature fitting automatically performed and based on the analytical thermal model of heat transfer in non-adiabatic conditions matches extremely well the experimental data.

The test was repeated 6 times for the brick and 3 times for the plaster. Results are reported in Table 2 and 3. It is worth noting that in both cases the test is repeatable with a precision about 2%.

Table 2. diffusivity measurements of a 5.5 mm thick brick sample by Non Linear Best Fitting

Test number	RMS	Diffusivity [10^7 m^2 s^{-1}]	Q/h [K m^{-1}]	Bi
1	0,0004	4,999	1653	0,172
2	0,0002	4,920	2131	0,123
3	0,0032	4,998	2379	0,121
4	0,0007	5,011	2444	0,224
5	0,0002	4,994	3439	0,231
6	0,0024	4,971	2420	0,191

Table 3. diffusivity measurements of a 5.6 mm thick plaster sample by Non Linear Best Fitting

Test number	RMS	Diffusivity [10^7 m^2 s^{-1}]	Q/h [K m^{-1}]	*Bi*
1	0.00121	6,819	655.392	0.209
2	0.00076	6,864	2077.72	0.166
3	0.00194	6,887	3177.49	0.115

Figure 16a shows a fragment of Attic ceramic (V° century B.C.) marked in few points for the sake of clarity of the following discussion. Figure16b illustrates the specimen inserted into the sample holder for the test. Figure 18a reports the diffusivity measured on the selected points. Here, the main problem is the non-planar and non-regular thickness of the sample. For this reason, a numerical simulation has been carried out in order to evaluate if differences in the experimental values are due to materials or the shape. Such a computation allowed drawing profiles of the temperature expected in ideal ant actual geometry. Figure 18b shows relative errors due to differentiating from the ideal flat and parallel surface, in the case of bended and irregular thickness measured on the real surface. The resulting maximum error is lower than 10% and actually happens for the point #5 and at minor extent for #2, #1.

Others more sophisticated techniques to measure thermal diffusivity has been developed and used for cultural heritage, having the possibility to perform measurements even in situ, without sampling the body [51].

3.2. In Situ Measurement of Thermal Inertia (Effusivity)

The importance of heat and mass transfer is a matter of facts dealing with the preservation of historical buildings.

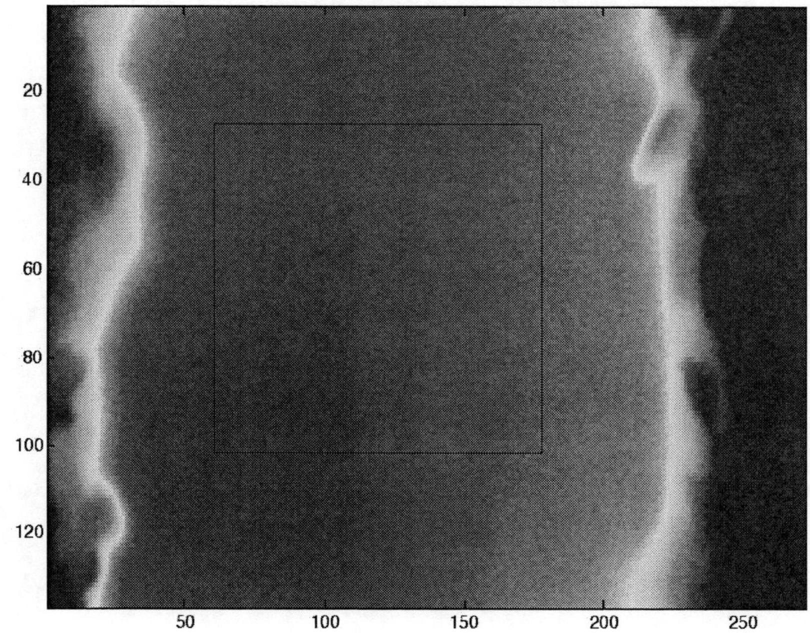

Figure15a. thermogram extracted from the sequence for the diffusivity measurement of a historical brick sample (not regular shape).

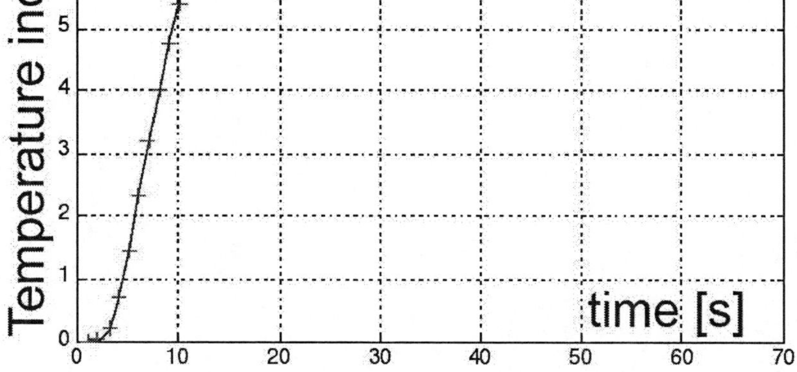

Figure 15b. Rear side average temperature history on the brick sample of fig16a (inside the box) and analytical function (solid curve).

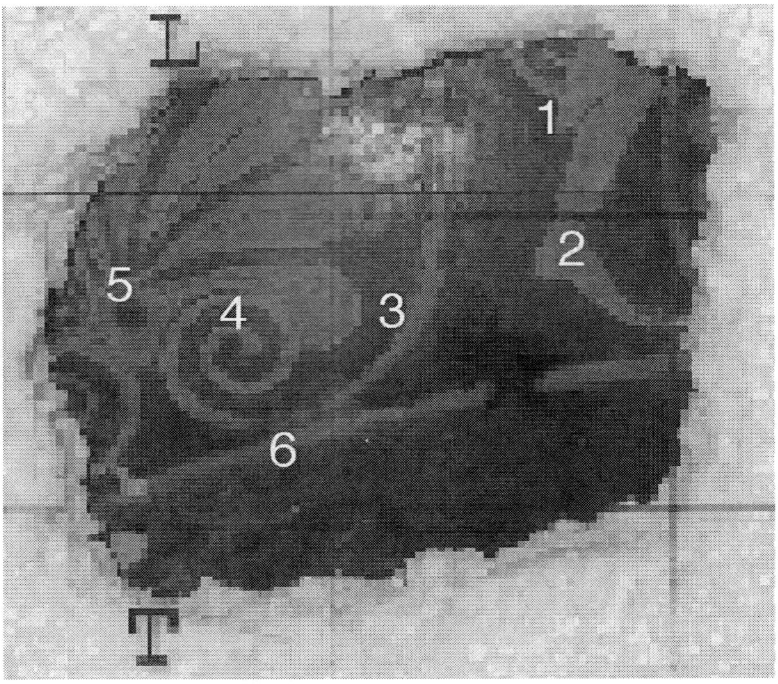

Figure 16a. Attic ceramic fragment (V° century B.C.) and selected points with different colour.

Figure 16b. specimen inserted into the sample holder with the heating device for the test

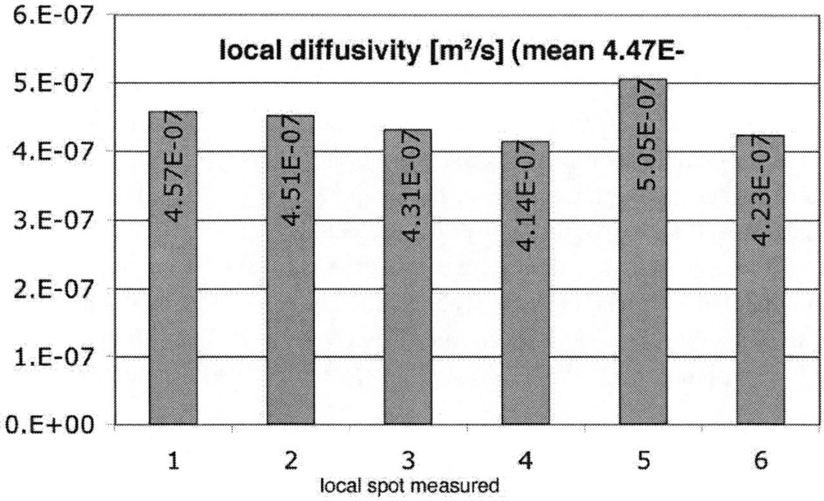

Figure 17a. Diffusivity measured on points marked in Figure17a (mean value 4.47 10^{-7} m^2 s^{-1}).

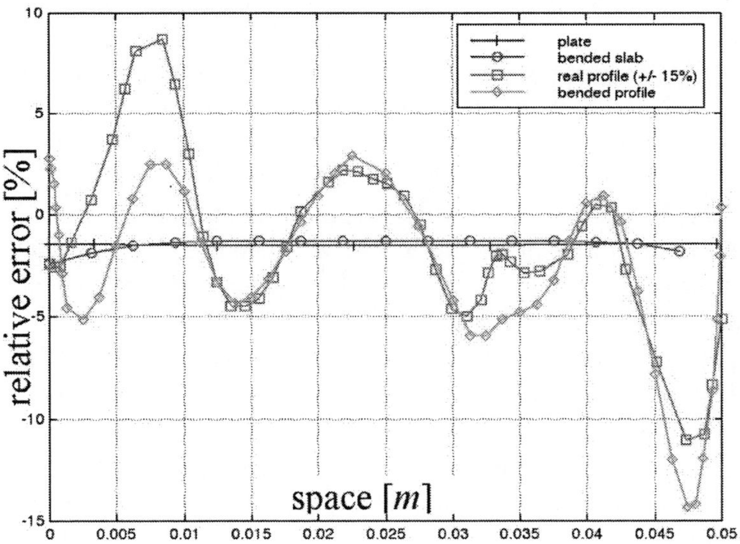

Figure 17b. Relative errors computed for the real thickness variation and bending of the sample.

It is well known that using of IR thermography could be very effective for the monitoring of the thermal stress due to solar radiation, radiative cooling and air temperature. The surface temperature map is a very important starting point, especially if it can be measured in different conditions over surfaces tilted and covered by dust or biological pollutants, as it happens for open-air monument.

An extensive temperature scanning of a monument by IR thermography in different micro-climatic conditions is a fast and very informative way to get useful data. These data contribute significantly to complete a mainframe needed for a correct restoration activity. Nevertheless, a more sophisticated approach allows measuring physical quantities that are directly correlated to the monument conservation status. In particular, the non-destructive testing of the monument by thermal method allows quickly identifying critical problems, as the material loss.

The addressed key-parameter for the effective analysis in situ of many pathologies is the Thermal Inertia or Effusivity (*e*). Thermal inertia, is the property of the material, that measures the temperature variation as reaction to the delivery of a certain amount of energy. The greater the thermal inertia, the lower the temperature variation. Thermal inertia is defined as the square root of density multiply by specific heat and conductivity. All of these quantities are generally decreased by the increase of the material porosity, due to the decay. Furthermore, if the plaster layer is not well attached to the background, the apparent effusivity sharply decreases, because the background is fairly isolated from the plaster. Vice versa, if the water content increases over the physiological value, effusivity increases proportionally.

Thermal inertia is expressed in SI units as $e = \sqrt{\lambda C \rho}$ [J m^{-2} K^{-1} $s^{-0.5}$]. That implies measurements of energy, space, temperature and time. Comparing thermal inertia with the thermal diffusivity, whose SI units are m^2 s^{-1} we notice how the latter needs only space and time measurements, being in that respect a simpler quantity to evaluate. It is obvious that the knowledge of both these quantities gives the thermal conductivity and the heat capacity of the material under test. It is worth noticing that thermal camera, if properly calibrated, is capable to measure not only temperature, but also time and space, because the imaging device is precisely timed.

Thermal inertia is a parameter that plays its role in the transient heating and cooling of bodies. From this point of view it differs from the heat capacity in so far the latter is measured in equilibrium conditions, while the former take into account also the ability of the body to diffuse the heat (via the thermal conductivity) not only to store it.

Thermal inertia, is measured in transient condition by imposing a heat flux on the surface of the material. That is principally realized by continuous, flash or eventually modulated lamps whose electromagnetic flux is transformed in heat once absorbed by the surfaces of the opaque bodies. Such transient techniques that utilize light sources are categorized as photothermal [52]. It is common to evaluate the transient temperature of the bodies by IR detector, minimizing any perturbation of the measured surface. Finally, suitable mathematical models gives the result.

The main purpose of the ongoing research is to apply the effusivity measurement in situ using as heat source the solar energy, with the purpose of materials characterization and monitoring the conservation status.

Thermal inertia measurement by optical means is quite a common practice in remote sensing application. Images collected by satellite in several bands of the visible and infrared spectrum are analyzed in order to estimate soil chararacteristics and moisture, mainly in meteorological and climatological application.

Generally, the thermal effect of the solar radiation on a monument is evaluated in a very complex way, using a model of the solar radiation damping due to the atmosphere; the angle effect, etc.. The energy injected on the monument surface depends on multiple aspects as: geography, topography and morphology; the heat and mass exchange with the surrounding; the absorption coefficient of the surface. Actually, several causes, e.g the carbon particles deposition from traffic and creating black crusts on monuments surface or the biological growth promoted by moisture increases significantly the absorption of the solar radiation and therefore the surface temperature.

Figure 6 reports an example of such an approach, using a simple linear model of the temperature variation vs. square root of time and considering constant the amplitude of the absorbed energy for the test period. A sharp step of the heat flux happens, when the shadow moves over the surface and this happens at different time. Therefore, a synchronisation is needed for different points.

Another example of such a monitoring is given in the following test performed on a church located close to Turin, Italy. The test has been run during the spring 2009 between 9:30 and 16:30.

In order to achieve the thermal inertia it must be measured the temperature increasing due to the energy input, according to a mathematical model, that takes into account the time dependance of the input (solar energy) and output (temperature). This technique is addressed locally for any point in a multiple steps procedure:

1) a preliminary spectral and directional evaluation of optical properties of the surface in order to simplify the further measure of the absorption coefficient;
2) a reflectance panoramic mapping by means of an imaging device accordingly to the peculiar characteristic of the materials; a global value is evaluated for each point by means of a radiometric reference placed in the field of view of the dedicated camera used and a calibration bar.
3) a solar radiometer measures the global irradiation on the analyzed plane;
4) temperature is measured by thermography at regular time steps in order to follow the transient thermal process;
5) data are processed by the mathematical model of equation (4) to compute the local effusivity value;
6) the resulting map is projected on the 3D geometrical model of the monument, in order to consider effects of shadows.

Accurate surface temperature is computed starting form the IR radiation detected by the thermal camera, corrected and calibrated through the radiometric reference indicated in Figure 19a. Figure 19b reports the mean temperature measured on the areas marked on Figure 19a and the air temperature, during the day.

At the same time, the absorbed solar energy (Q) is given multiplying the global irradiation, times the absorption coefficient of any point. Irradiation is measured on the wall plane by the solar radiometer indicated in Figure 18a. Figure 18b shows also the solar irradiation (magenta curve) in terms of $W\ m^{-2}$. Because of favourable characteristic of building materials in terms of reflectivity and transparency, a simplified technique for measuring the absorption coefficient has been set up [53]. The basic idea is to limit the detected radiation to visible and near IR bands, using a modified camera and a calibration bar (manufactured by Kodak®). Figure 18c shows the map of absorption coefficient measured for the inspected area and registered on the thermographic map.

This allows making a balance of the heat flux, which gives a local value of the thermal inertia of any point of the surface according to eq. (4).

$$T_t - T_0 = \frac{Q}{h} A_0 \sin(\omega t - \Psi) \qquad (4)$$

where ω is the solar angle (related to true solar time) A_0 the amplitude and Ψ the phase angle:

$$A_0 = \left(1 + \sqrt{2}\frac{\sqrt{\omega}}{h}e_{x,y} + \left(\frac{\sqrt{\omega}}{h}e_{x,y}\right)^2\right)^{-1/2} \quad \text{e} \quad \Psi = \arctan\left(\frac{\frac{\sqrt{\omega}}{h}e_{x,y}}{\sqrt{2} + \frac{\sqrt{\omega}}{h}e_{x,y}}\right)$$

Finally the resulting map of effusivity is geometrically corrected and placed on the geo-referenced 3D model of the monument, as seen in Figure 18d.

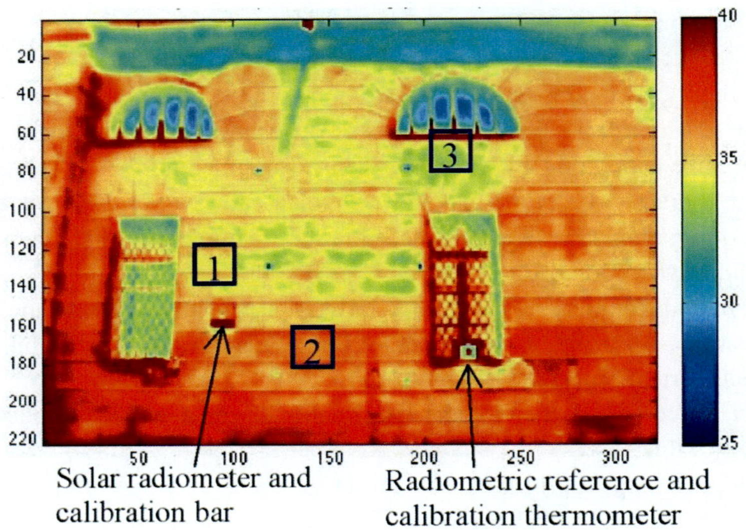

Figure 18a. Raw thermogram (with perspective distortion) taken at 12:30 during the test.

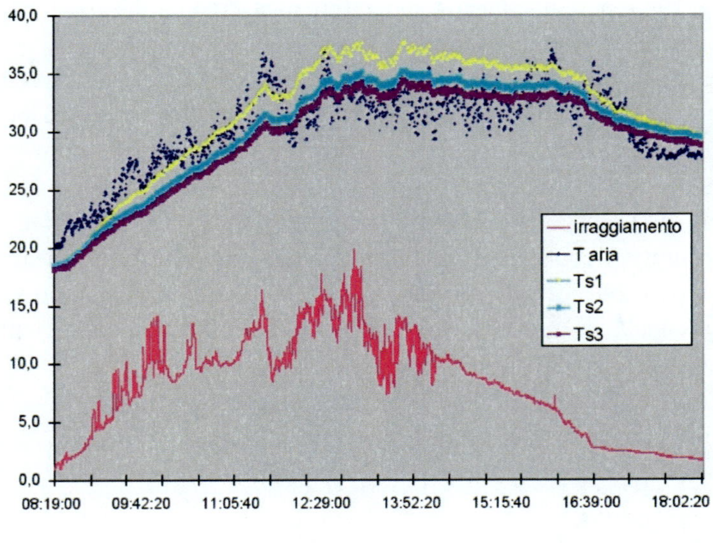

Figure 18b. Surface temperature vs. time at selected points, as seen in Figure 19a.

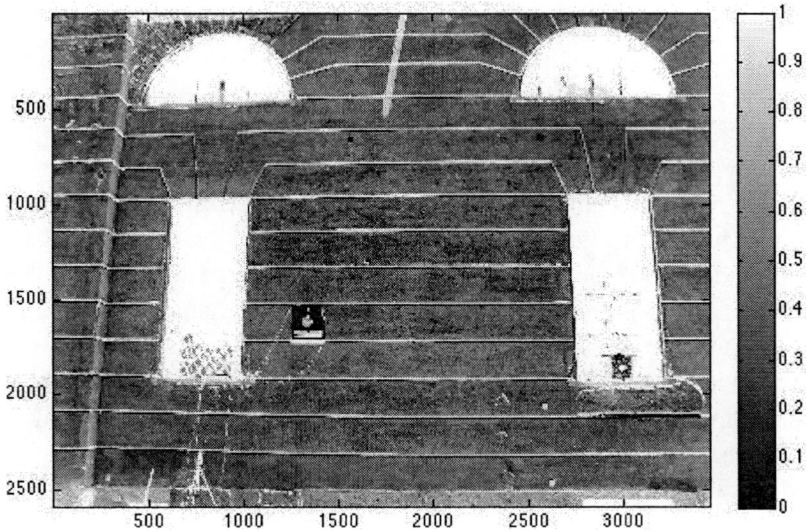

Figure 18c. map of the absorption coefficient of the surface.

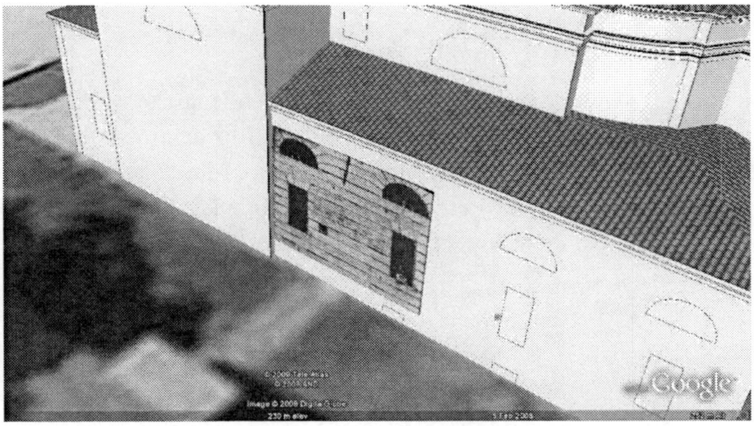

Figure 18d. Effusivity map over the 3D model of the church, including shadows and orientation.

CONCLUSION

The applications of IR Thermography for cultural heritage conservation is greatly improved by the usage of mathematical model of heat and mass transfer. Particularly, the non destructive testing and evaluation of materials are much more reliable when supported by a test simulation for the optimisation of experimental parameters and processing algorithms. This approach improves the effectiveness of the test, reducing the cost due to vane attempts and fortuitous repetitions by chance and allows verifying different hypothesis about the target. The final result is the geometric characterization of searched discontinuity, in the three dimensions.

Various kind of tests have been reported, making thermographic inspection more reliable, practical and able to be performed both in laboratory and in situ.

When dealing with building materials it is common to encounter a lack of information about the thermal properties. On the other hand, being these materials generally multi-layered, the knowledge of each layer properties is crucial for the solution of the inverse problem. Thermography allows setting up effective techniques to characterize such a precious materials.

Researches of moisture content in buildings of artistic interest, carried out for restoration activities, take advantage of the imaging technique in the visible and infrared bands, both for their non-intrusiveness, and for the possibility of scanning large surfaces in a small time. A new technique is also presented capable to monitor the boundary conditions with high accuracy and resolution.

The adopted equipment is specially set up in order to fit particular requirements of the building inspection and to manage the huge amount of stored data, especially for transient techniques when the analysis is performed both in space and time domain.

Presented case studies are a small selection among many others results of applied researches, all demonstrating the great capability of the thermal method for the non destructive investigation of the world around us.

ACKNOWLEDGMENTS

Author is highly in debt with many friends and colleagues working together on the presented research projects. Among others I would like to acknowledge the precious help given by Paolo Bison, Chiara Bressan, Gianluca Cadelano, Mara Gava, Mario Girotto, Sergio Marinetti, Andrea Mazzoldi, Fabio Peron, Elisabetta Rosina, Ayse Tavukcuoğlu, Vladimir Vavilov, Monica Volinia and their important contribution.

REFERENCES

[1] Maldague, X., *Theory and Practice of Infrared Technology for NonDestructive Testing*, John Wiley-Interscience, 2001.

[2] ASNT Nondestructive Testing Handbook, *Infrared and Thermal Testing*. third edition: Volume 3,Technical Editor: Xavier P.V. Maldague. Editor: Patrick O. Moore, 2001.

[3] Grinzato, E.; Bison, P.G.; Marinetti, S. *J. Cultural Heritage* 2002, *3*, 21-29.

[4] Krapez, J.C.; Balageas, D. In Proc. of the Eurotherm Seminar 42 *Quantitative Infrared Thermography-QIRT'94*, Aug.23-26, 1994, Sorrento, Italy, pp.260-266.

[5] Wu, D.; Salerno, A.; Schnbach, B.; Halin, H.; Busse, G. In Proc. SPIE 3056 *Thermosense-XIX*, 1997, p.176-183.

[6] Vavilov, V.; Bison, P.G.; Bressan, C.; Grinzato, E.; Marinetti, S. *Soviet Journal Non-Destructive Testing* Plenum Publishing Co. (USA), 1061-8309/92/2807-0419, pp. 419-424.

[7] Crowther, D.J.; Favro, L.D.; Kuo, P.K.; Thomas R.L. *J.Appl.Phys*. 1993, *74*, 5828-5834.

[8] Aamodt, L.C.; Maclachlan Spicer J.W.; Murphy, J.C. *J. Appl. Phys*. 1990, *68*, 6087-6098.

[9] Vavilov, V.; Kauppinen, T.; Grinzato E. *Res. Nondestr. Eval.* 1997, *9*, 181-200.
[10] Grinzato, E.; Bison, P.G.; Marinetti, S.; Vavilov, V. *Insight* 2001, *43*, 254-259.
[11] Carslaw, H.S.; Jaeger J.C., *Conduction of heat in solids*. Oxford University Press. 1959
[12] Lau, S.K.; Almond, D.P.; Milne J.M.. *NDTand E International* 1991, *24*, 195-202.
[13] Balageas, D.L.; Krapez, J.C.; Cielo, P. *J. Appl. Phys.* 1986, *59* 348-357.
[14] Parker, W.J.; Jenkins, R.J.; Butler, C.P.; Abbott, G.L. *J. Appl. Phys.* 1961, *32*, 1679-1684.
[15] Busse, G.; Wu, D.; Karpen, W. *J. Appl. Phys.* 1992, *71*, 3962-3965.
[16] Gruss, C.; Balageas, D. In Proc. *QIRT 92*-Eurotherm Series 27-EETI ed., Paris 1992, pp. 19-24.
[17] Bison, P.G.; Grinzato, E.; Marinetti, S. In Proc. *QIRT Journal* Lavoisier ed.2004, *1*, 241-250.
[18] Fabbri, L.; Fenici, P. *Rev. Sci. Instrum* 1995, *66*, 3593-3600.
[19] Moussa, M.; Maldague, X. In Proc. Workshop on *Advanced infrared technology and applications*, September 19-20, 1995 Capri (Italy), Fondazione G. Ronchi
[20] Cramer, K.E.; Winfree, W.P.; Howell, P.A.; Syed, H.; Renouard, K.A. In Proc. SPIE *Thermosense XIV* 1992, vol. 1682, pp. 162-170.
[21] Vavilov, V.; Kourtenkov, D.G.; Grinzato, E.; Bison, P.G.; Marinetti, S.; Bressan C. In Proc. *Eurotherm Seminar n. 42, QIRT 94* D. Balageas, G. Busse and G.M. Carlomagno Eds. EETI editions, Paris, France, pp.273-278, Sorrento (Italy), 1994.
[22] Marinetti, S.; Muscio, A.; Bison, P.G.; Grinzato, E. In Proc. SPIE *Thermosense XXII* 2000, vol. 4020, pp. 164-173.
[23] Grinzato, E.; Rosina, E.; Vavilov, V.; Grimoldi, A.; Di Flumeri, E.; Papetti, V.; Roche, G. 3rd International Congress on *Science and Technology for the Safeguard of Cultural Heritage in the Mediterranean Basin*, p.393, Halcalà de Henares (2001).
[24] Ljungberg, S.A. *Infrared Methodology and Technology* X. Maldague Ed., Gordon and Breach Science Publishers, 1994 vol.7, Chap. 6 pp. 211-252.
[25] Grinzato, E.; Vavilov, V.; Kauppinen, T. *Energy and Buildings* 1998, 29, 1-9.
[26] Grinzato, E.; Vavilov, V.; Marinetti, S.; Bison, P.G. In Proc. 10° Congresso nazionale *AIPnD*, pp.81-90, Ravenna 2003.
[27] Grinzato, E.; Bressan, C.; Peron, F.; Romagnoni, P.; Stevan, A.G. In Proc. SPIE *Thermosense XXII* 2000, vol.4020, pp. 314-323.
[28] Grinzato, E. *Infrared Methodology and Technology*, X. Maldague Ed.; Gordon and Breach Science Publisher, 1994, vol. 7, Chap.4, pp.103-128.
[29] Vavilov, V. In Proc SPIE *Thermosense XII*, 1990, vol. 1313, pp. 178-182.
[30] Vavilov, V.; Bison, P.G.; Bressan, C.; Grinzato, E.; Marinetti, S. In Proc. *Quantitative InfraRed Thermography QIRT 92*, Editions Europeennes Thermique et Industrie, 1992, pp.259-265.
[31] Maldague, X.; Marinetti, S. *J.Appl.Phys.* 1996, *79*, 2694-2698.
[32] Rajic, N. *Composite Structures* 2002, *58*, 521–528.
[33] Shepard, S.M. In Proc. SPIE *Thermosense XXIII* 2001, vol. 4360, pp. 511-515.
[34] Grinzato, E.; Bison, P.G.; Marinetti, S.; Vavilov, V. *Research in Nondestructive Evaluation* 1994, *5*, .257-271.
[35] Ibarra-Castagnedo, C.; Maldague, X. *QIRT Journal* 2004, *1*, 47-70.
[36] Marinetti, S.; Grinzato, E.; Bison, P.G.; Bozzi, E.; Chimenti, M.; Pieri, G.; Salvetti, O. *Journal of Infrared Physics and Technology* 2004 *46*, 85-91.

[37] Grinzato, E.; Bison, P.G., Bressan, C.; Mazzoldi, A.; Danti, C.; Vavilov, V. In Proc. Workshop on *Advanced Infrared Technology and Applications* Capri (Italy), 1995, pp.273-288
[38] Grinzato, E.; Bison, P.G.; Bressan, C.; Mazzoldi, A. In Proc. *QIRT 98* Eurotherm Seminar n. 60, Lodz (Poland), 1998, pp.64-67.
[39] Grinzato E.; Cadelano, G.; Bison, P.; Petracca, A. In Proc. SPIE *Thermosense XXXI*, 2009, vol. 7299,
[40] Debevec, P.E.; Mali, J. Recovering High Dynamic Range Radiance Maps from Photographs
[41] Grinzato, E.; Bressan, C.; Bison, P.G.; Mazzoldi, A.; Baggio, P.; Bonacina, C. In Proc. SPIE *Thermosense XIV* 1992 ,vol. 1682, pp. 213-221
[42] Baggio, P.; Bison, P.G.; Bonacina, C.; Bressan, C.; Grinzato, E. In Proc. *XVIII International Congress of Refrigeration*. August 10-17, 1991 Montreal, Quebec, Canada.
[43] Tavukcuoglu, A.; Cicek, P.; Grinzato, E. *Quantitative InfraRed Thermography Journal (QIRT)* 2008, *5*, 151-174.
[44] Tavukcuoğlu, A.; Düzgüneş, A.; Demirci, Ş.; Caner-Saltık, E.N. *Building and Environment*, 2007, *42*, 2699–2709.
[45] Griffin, C.W.; Fricklas, R.L. Draining the roof. Manual of low-slope roof systems, 3rd ed. New York: McGraw Hill, 1995.
[46] Grinzato, E.; Peron, F.; Strada, M. In Proc SPIE *Thermosense XXI* 1999, *3700*, pp. 471-482.
[47] Pretto, A.; Menegatti, E.; Bison, P.; Grinzato, E. In Proceedings *NDT in Canada*, London (Canada), 2009 August 27-28.
[48] Grinzato, E. *A method and apparatus for thermal-hygrometric monitoring of wide surfaces by IR Thermography* 2006, n° PD2006A000191.
[49] Bison, P.G.; Marinetti, S.; Mazzoldi, A.; Grinzato, E.; Bressan, C. *Journal of Infrared Physics and Technology* 2002, *43*, 127-132.
[50] Grinzato, E.; Marinetti, S. *Advances in signal processing for non destructive evaluation of materials*, ed. by X. Maldague, NATO ASI, Series E: Applied Sciences, Kluwer Academic Publishers, (1994) vol.262. pp.117-132.
[51] Bison, P.G.; Cernuschi, F.; Grinzato, E. *Int. J. Thermophysics* 2008, *29*, 2149-2161.
[52] Almond, D.P.; Patel, P.M. *Photothermal Science and Techniques*. Chapman and Hall 1996.
[53] Grinzato, E.; Bison, P.; Girotto, M.; Volinia, M. In Proc. Conferenza Nazionale *AIPnD*, Roma 2009 (www.ndt.net/articles/aipnd2009/files)

In: Recent Advances in Non-Destructive Inspection
Editor: Carosena Meola, pp. 161-190
ISBN 978-1-61668-550-8

Chapter 6

ELECTROMAGNETIC TECHNIQUES AND HIGH SENSITIVITY MAGNETIC SENSORS IN NDT FOR AERONAUTICAL APPLICATIONS

Massimo Valentino and Carmela Bonavolontà
CNR-INFM Coherentia, Piazzale V. Tecchio 80, 80125 Napoli, Italy

ABSTRACT

This work is a comparative study of non-destructive testing performed by different sensors: search coil, flux-gate, GMR, HTc-SQUID-based magnetometer, used as eddy-current measuring devices. These sensors have been used for the detection of various artificial defects affecting metallic multi-layered, aluminium-alloy structures and composite materials, without external electromagnetic shielding. The basic concepts of eddy current technique and the working principles of the most useful electromagnetic sensors are reported. In particular, we demonstrate that the higher sensitivity of SQUIDs, in comparison to the other electromagnetic probes, gives a decisive advantage when the structure under inspection are multi-layered or weakly conducting materials such as the new advanced composite materials like CFRP (Carbon Fiber Reinforced Plastic), and Fiber Metallic Laminate (FML).

INTRODUCTION

Characterisation of material properties is critical for understanding the mechanical behaviour and design performance under its operating conditions. Tailoring the material properties, which are functions of the material states, is essential for advanced product design. The necessity to characterise materials for a myriad of applications has spurred the development of many new methods and instruments. However, many of these tools require destructive sectioning of the material to provide characterisation. Therefore, many of these techniques do not provide key information about material parameters in their operating environments. An ideal characterisation tool would provide data about the material properties that affect micro- and macro- structural behaviour without the need to section destructively the sample. Such data can be obtained only using Non Destructive Evaluation (NDE) methodologies.

NDE has become an integral part of material research and the last years have been marked with the most incredible progress. The majority of the current NDE methods were introduced around the middle of the last century or before ,i.e. the eddy-current testing, and the followed modifications, improvements and enhancement contributed to an unprecedented advancement in capability and reliability. The requirements for NDE are driven by the need for lower cost methods and instruments with greater reliability, sensitivity, user friendliness and high operation speed as well as applicability to new and complex materials, such as composites, and structures with complex geometries such as aircraft, bridges, pipelines, and pressure vessel. Nowadays, NDE research covers, as interdisciplinary field, many scientific disciplines benefited from capabilities that were developed in many other branches of science and engineering as Physics, Material Science and Electronic, to improve the NDE technique, leading to smarter instruments, which are computer controlled, smaller, lighter and more capable.

In the industry, NDE is used in process control, in post-production quality control, and in the testing of system that are already in use. It is concerned in a practical way with the performance of the test piece, it summarize the following question: "how long may the piece be used and when does it need to be checked again?" Non-destructive testing in the industry has become an increasingly vital factor in the effective conduct of research, development, design and manufacturing programs. Only with appropriate use of non-destructive testing techniques can the benefits of advanced materials science be fully realized.

The objectives of non-destructive testing methods, as suggested by the National Materials Advisory Board (NMAB), ad hoc Committee on Non-destructive Evaluation, are defined in Table 1. In the latter the first column include the goal of the detection, while the column on the right include a brief description of the physical characteristic of the anomalies that can be detected using electromagnetic techniques. In the following the eddy current testing technique will be presented and several testing reported in the table 1 will be discussed.

In this work the section 2 is dedicate to the working principle of Eddy Current technique, while the section 3 reports the description of the most exploited electromagnetic sensors such as induction coil, SQUIDs (Superconducting Interference Devices), GMR (Giant Magneto-Resistance) and Flux gate.

In the following the eddy current and the working principles of the sensors are summarized in the paragraph 2 and 3 respectively. To understand the drawbacks and advantages that each sensor offers, a comparison in terms of magnetic field sensitivity, noise spectral density and frequency dependence parameters is reported in the paragraphs 4-7. To emphasise the capability of eddy current technique based on magnetic field sensors, experimental results related to inspection of aeronautical metallic and composite structures are reported. In particular, experimental data concerning the detection of surface and sub-surface flaw in multy-layered metallic structures have been shown in the section 8, while section 9 and 10 regard the application of eddy current technique based on electromagnetic sensors to detect internal defects into complex and advanced composite materials such as FML (Fiber Metal Laminate) and CFRPs (Carbon Fiber Reinforced Plastics).

Table 1. Objectives of non-destructive testing methods using eddy current technique

Objectives	Attributes Measured or Detected
Discontinuities	colour, cracks, dimensions, film thickness, gaging, reflectivity, strain distribution and magnitude, surface finish, surface flaws, through-cracks
Surface anomalies	roughness, scratches, gouges, crazing, pitting, inclusions and imbedded foreign material
Surface connected anomalies	cracks, porosity, pinholes, laps, seams, folds, inclusions
Internal anomalies	cracks, separations, hot tears, cold shuts, shrinkage, voids, lack of fusion, pores, cavities, delaminations, disbonds, poor bonds, inclusions, segregations
Structure	
Microstructure	molecular structure, crystalline structure and/or strain, lattice structure, strain, dislocation, vacancy, deformation
Matrix structure	grain structure, size, orientation and phase, sinter and porosity, impregnation, filler and/or reinforcement distribution, anisotropy, heterogeneity, segregation
Small structural anomalies	leaks (lack of seal or through-holes), poor fit, poor contact, loose parts, loose particles, foreign objects
Gross structural anomalies	assembly errors, misalignment, poor spacing or ordering, deformation, malformation, missing parts
Dimensions and metrology	
Displacement, position	linear measurement, separation, gap size, discontinuity size, depth, location and orientation
Dimensional variations	unevenness, non-uniformity, eccentricity, shape and contour, size and mass variations
Thickness, density	film, coating, layer, plating, wall and sheet thickness, density or thickness variations
Physical and mechanical properties	
Electrical properties	resistivity, conductivity, dielectric constant and dissipation factor
Magnetic properties	polarization, permeability, ferromagnetism, cohesive force
Thermal properties	conductivity, thermal time constant and thermoelectric potential
Mechanical properties	compressive, shear and tensile strength (and module), Poisson's ratio, sonic velocity, hardness, temper and embrittlement
Surface properties	colour, reflectivity, refraction index, emissivity
Chemical composition and analysis	
Elemental analysis	detection, identification, distribution and/or profile
Impurity concentrations	contamination, depletion and doping
Metallurgical content	variation, alloy identification, verification and sorting
Physiochemical state	moisture content, degree of cure, ion concentrations and corrosion, reaction products

Table 1. (Continued)

Objectives	Attributes Measured or Detected
Stress and dynamic response	
Stress, strain, fatigue	heat-treatment, annealing and cold-work effects, residual stress and strain, fatigue damage and life (residual)
Mechanical damage	wear, spalling, erosion, friction effects
Chemical damage	corrosion, stress corrosion, phase transformation
Other damage	radiation damage and high frequency voltage breakdown
Dynamic performance	crack initiation and propagation, plastic deformation, creep, excessive motion, vibration, damping, timing of events, any anomalous behaviour
Signature analysis	
Electromagnetic field	potential, strength, field distribution and pattern
Signal or image analysis	image enhancement and quantization, pattern recognition, densitometry, signal classification, separation and correlation, discontinuity identification, definition (size and shape) and distribution analysis, discontinuity mapping and display

1. Histrorical Information

Eddy current testing has its origins with Michael Faraday's discovery of electromagnetic induction in 1831. Faraday was a scientist in England during the early 1800's and is credited with the discovery of electromagnetic induction, electromagnetic rotations, the magneto-optical effect, diamagnetism, and other phenomena.

Few years later the French physicist Leon Foucault in 1851, built a device that used a copper disk moving in a strong magnetic field to show that eddy currents (magnetic fields) are generated when a material moves within an applied magnetic field. For this reason eddy currents are sometimes called Foucault currents.

In 1879, another scientist named Hughes recorded changes in the properties of a coil when placed in contact with metals of different conductivity and permeability.

Beginning in 1933, in Germany, Professor Friedrich Förster adapted eddy current technology to industrial use, developing instruments for measuring conductivity and for sorting mixed-up ferrous components. Many advances were made throughout the 1950s and 1960s, especially in the aircraft and nuclear industries. There have been many recent developments in eddy current testing, leading to improved performance and the development of new applications. Eddy current testing is now a widely used and well-understood inspection technique for flaw detection as well as for thickness and conductivity measurements. Eddy current has been one of the leading in-service inspection methods for crack detection around and inside fastener holes.

Eddy current inspection is one of several NDT methods that use the principles of electromagnetic induction. When an alternating current is used to feed a coil, an alternating magnetic field is produced and the magnetic flux lines are concentrated at the center of the coil. Then, when this coil is brought near an electrically conductive material, the alternating magnetic field penetrates the material and generates continuous, circular eddy currents as schematized in figure 1 up. Larger eddy current are produced near the test surface and as the penetration of the induced field increases, the eddy current become weaker. The induced eddy current produces an opposing (secondary) magnetic field in the opposite direction to the generated (primary) magnetic field. This opposing magnetic field, coming from the material, has a weakening effect on the primary magnetic field and the test coil can sense this change. Eddy currents are closed loops of induced current circulating in planes perpendicular to the magnetic flux. They normally travel parallel to the coil's winding and flow is limited to the area of the inducing magnetic field. Eddy currents concentrate near the surface adjacent to an excitation coil and their strength decreases with distance from the coil.

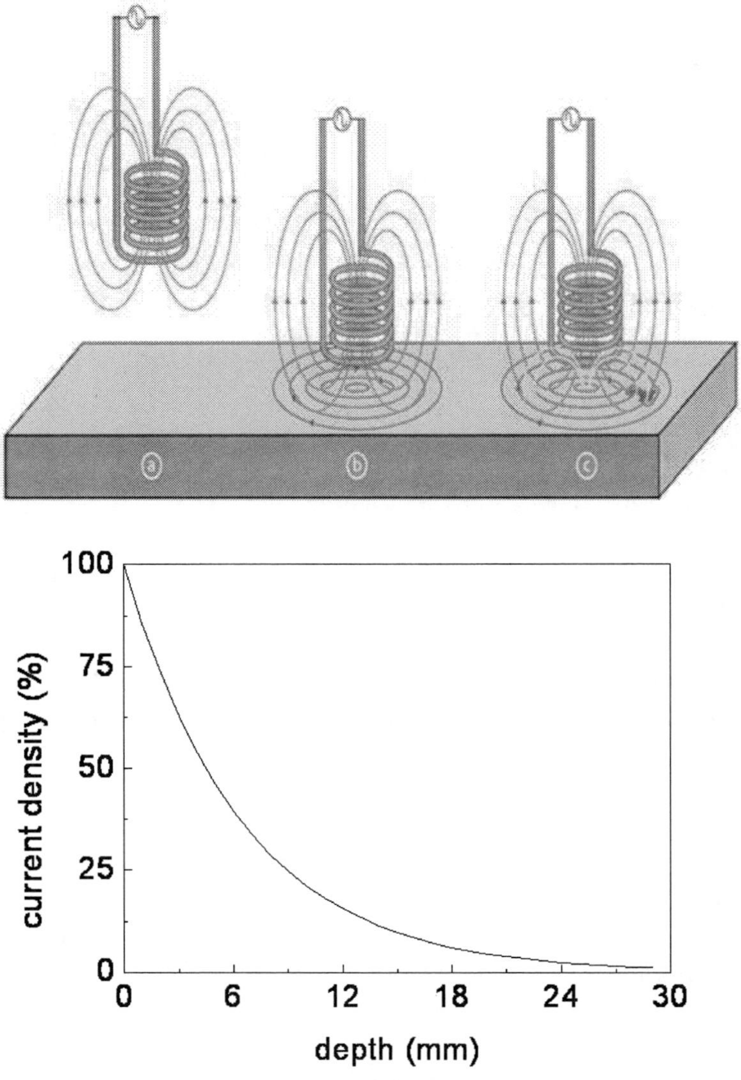

Figure 1. Image of the eddy current distribution produced by an solenoid coil. (a) no contact probe, (b) no defect detection, (c) defect detection, in the picture above. The density of the eddy current at different depth in a steel sample at frequency of 277 Hz, below.

In effect, the impedance of the coil is reduced proportionally as the eddy current are increased in the test piece. A crack in the test material obstructs the eddy current flow, lengthens the eddy current path, reduces the secondary magnetic field, and increases the coil impedance. The inductive reactance of the coil continues to increase as the severity of the defect increases. The modelling of flaws response contribute significantly to the understanding of the key parameters as well as the reduction of noise and lift-off.

Eddy current density decreases exponentially with depth inside a material, as shown in the figure 1(down). This phenomenon is known as the skin effect. Skin effect arises when the eddy currents flowing in the test object at any depth produce magnetic fields, which oppose the primary field, thus reducing net magnetic flux and causing a decrease in current flow as depth increases. Alternatively, eddy currents near the surface can be viewed as shielding the coil's magnetic field, thereby weakening the magnetic field at greater depths and reducing induced currents.

The depth that eddy currents penetrate into a material depends on the frequency of the excitation current, the electrical conductivity and the magnetic permeability of the specimen. The depth of penetration decreases with increasing frequency and increasing conductivity and magnetic permeability. The depth at which eddy current density has decreased to $1/e$, or about 37% of the surface density, is called the standard depth of penetration (δ), usually it is defined as:

$$\delta = \frac{1}{\sqrt{\pi \sigma f \mu}} \tag{1}$$

δ = Standard Depth of Penetration (mm)
f = Test Frequency (Hz)
μ = Magnetic Permeability (H/mm)
σ = Electrical Conductivity (% IACS or S/m)

The word 'standard' denotes plane wave electromagnetic field excitation within the test sample (conditions which are rarely achieved in practice). Although eddy currents penetrate deeper than one standard depth of penetration they decrease rapidly with depth. At two standard depths of penetration (2δ), eddy current density has decreased to $1/e$ squared or 13.5% of the surface density. At three depths (3δ) the eddy current density is down to only 5% of the surface density. The electromagnetic skin depth versus the frequency and for different value of the electrical conductivity is shown in figure 2 for aluminium alloy of aeronautical interest.

Eddy current inspection is used in a variety of industrial applications to find defects and make measurements. One of the primary uses of eddy current testing is for defect detection when the nature of the defect is well understood. In general, the technique is used to inspect a relatively small area and the probe design and test parameters must be established with a good understanding of the flaw that is trying to be detected. Since eddy currents tend to concentrate at the surface of a material, they can only be used to detect surface and near surface defects as shown in Figure 2 for the case of aluminium alloy with different electrical conductivity. Moreover, enhancement of probes and instrumentation sensitivity has been allowed the effective use of eddy current at low frequencies to detect flaws in metallic multi-layered structures.

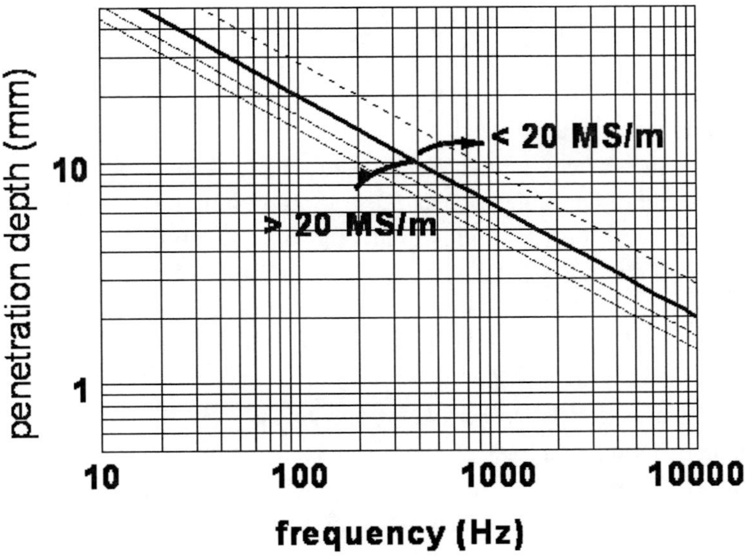

Figure 2. The skin depth for different aluminium alloy of aeronautical interest. Increasing the electrical conductivity of the material, the electromagnetic penetration depth reduces as seen above.

In thin materials such as tubing and sheet stock, eddy currents can be used to measure the thickness of the material. Moreover, the eddy current technique is used to make corrosion thinning measurements on aircraft skins and in the walls of tubing used in assemblies such as heat exchangers. Eddy current testing is also used to measure the thickness of paints and other coatings. Eddy currents are also affected by the electrical conductivity and magnetic permeability of materials. Therefore, eddy current measurements can be used to sort materials and to tell if a material has seen high temperatures or been heat treated, which changes the conductivity of some materials.

2. Probe and Sensors for Eddy Current Applications

2.1. Induction Coils

Eddy current probes are available in a large variety of shapes and sizes. In fact, one of the major advantages of eddy current inspection is that probes can be custom designed for a wide variety of applications. Eddy current probes are classified by the configuration and mode of operation of the test coils. The configuration of the probe generally refers to the way the coil or coils are packaged to best "couple" to the test area of interest. An example of different configurations of probes would be bobbin probes, which are inserted into a piece of pipe to inspect from the inside out, versus encircling probes, in which the coil or coils encircle the pipe to inspect from the outside in. The mode of operation refers to the way the coil or coils are wired and interface with the test equipment. The most widely used mode of operation of a probe is Absolute and Differential.

Absolute probes generally have a single test coil that is used to generate the eddy currents and sense changes in the eddy current field. By measuring the absolute change in impedance of the test coil, much information can be gained about the test material. Absolute coils can be used for flaw detection, conductivity measurements, lift-off measurements and thickness measurements. They are widely used due to their versatility. Since absolute probes are sensitivity to things such as conductivity, permeability lift-off and temperature, steps must be taken to minimize these variables when they are not important to the inspection being performed. It is very common for commercially available absolute probes to have a fixed "air loaded" reference coil that compensates for ambient temperature variations.

Differential probes have two active coils usually wound in opposition, although they could be wound in addition with similar results. When the two coils are over a flaw-free area of test sample, there is no differential signal developed between the coils since they are both inspecting identical material. However, when one coil is over a defect and the other is over good material, a differential signal is produced. They have the advantage of being very sensitive to defect yet relatively insensitive to slowly varying properties such as gradual dimensional or temperature variations. Probe wobble signals are also reduced with this probe type. There are also disadvantages to using differential probes. Most notably, the signals may be difficult to interpret. For example, if a flaw is longer than the spacing between the two coils, only the leading and trailing edges will be detected due to signal cancellation when both coils sense the flaw equally.

The experimental results shown below has been obtained using the absolute probe type PKA 22-6, [20] with a diameter of 38 mm and a thickness of 40 mm, at an operating frequency ranging from 50 Hz to 1 kHz. The maximum voltage sensitivity of this probe is 24 μV at 70 Hz. This particular kind of coil, optimized for operation at low frequency, has been chosen for a detection of deep flaws. It can be positioned on the surface of the specimen at a minimum distance z = 0.3 mm. The experimental results shown in the following have been obtained by moving the sample with a speed of 1 mm/s beneath the stationary probes.

2.2. SQUID Magnetometers

The heart of the SQUID technology is the Josephson junction that can be realized by means of a thin insulating layer or a narrow constriction between two superconductors. The electrical properties of the Josephson junction are described by two equations. The first equation related the applied current passing through the junction to the relative phase and the critical current of the junction I_C. The second equation relates the voltage across the junction to the temporal derivative of relative phase [12-13]. The magnetic features of the SQUID is fixed by the flux quantization. It asserts that the magnetic flux in a superconducting ring has to be integer units of the flux quantum (Φ_0 = 2.07 x 10^{-15} Wb). Because the latter magnetic property, the superconductive current which flow in the ring, holds the enclosed flux constant opposing any change in magnetic flux applied to the ring. Generally, the materials used in the SQUID fabrication are Niobium and the compound $YBa_2Cu_3O_{7-x}$. The first is a Low Temperature metallic Superconductor (LTS) with a critical temperature of 9.25 K. The second is a ceramic superconductor and it belongs to a class of High T_c Superconductors (HTS) characterized by a critical temperature of 93 K. Typically, the LTS SQUIDs are cooled by immersion in liquid helium which has a boiling point of 4.2 K, while the HTS SQUIDs can be operated in liquid nitrogen characterized by a boiling point of 77 K. In NDE applications the HTS SQUIDs are preferred because the liquid nitrogen is cheaper and easier to handle than the liquid helium.

SQUID is a highly sensitive magnetic field sensor (0.1pT /Hz $^{1/2}$ at 1 Hz) for eddy current measurements even in harsh environmental conditions. SQUIDs in general have a very high magnetic field sensitivity which is nearly independent of frequency from 0 Hz up to 1 MHz. This offers an advantage for magnetic field sensor for applications where a low excitation frequency especially for large depth of penetration is required. The device requires cooling to cryogenic levels, usually liquid Nitrogen, but nowadays criogenerator system with a good thermal stability are available. Recent research in SQUID based NDE systems has proven their superiority over conventional systems when searching for cracks depth 10 mm or more in *Al* alloy sample [9,11]. Further, it was shown to have an improved signal to noise ratio for cracks deeper than 11 mm as compared with traditional eddy current technique [14]. Additionally, the high bandwidth available with certain SQUID systems make measurements over a wide frequency range without having to change the sensor. The high dynamic range (the ratio between the highest field change which can be measured before the system goes into saturation and the lowest detectable field) allows one to detect small field changes in the presence of large background fields, produced by e.g. edge effects or inhomogeneities in conductivity. The method is contact less and slightly sensitive on the distance between the sample and the sensor (lift-off). The probe is effective mostly in low frequency applications for the detection of deep lying defects, in multilayer structures, rivet plates, aircraft wheels and rotor plates engine. The large dynamic range of SQUID allows shorter integration time and therefore faster scanning is possible (more than 5 cm/s).

The NDE eddy current system based on SQUID can been schematized as shown in Figure 3.

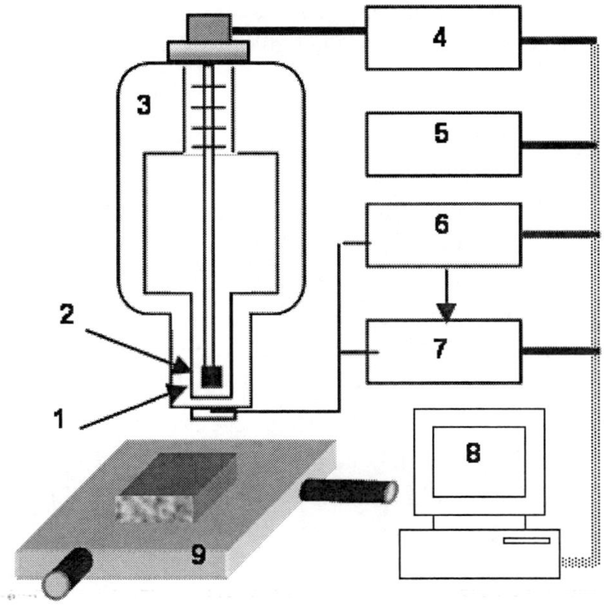

Figure 3. Typical scheme of the eddy-current SQUID-based NDE prototype. 1 excitation coil; 2 SQUID sensor; 3 cryogenic dewar; 4 SQUID electronics; 5 pre-amplifier electronics; 6 current supply; 7 lock in amplifier; 8 PC; 9 x-y scanning table.

The SQUID signal is synchronously demodulated by a dual-channel lock-in amplifier, locked at the frequency of the excitation current. The magnitude and phase of the magnetic field are recorded by a computer. A non-metallic and non-magnetic computer-controlled x-y positioning system moves the specimens under the cryostat with a speed ranging from 1 mm/s to 10 mm/s. The surface scans of the specimens were acquired in continuous mode with a typical rate of 10 points/mm.

The SQUID pick up coil is placed in the center of the excitation coil so that a zero field detector is realized. Two different excitation coil can be used: gradiometric (2D shape) and circular. The first is typically coupled with a SQUID pick up coil and it is sensible to change of vertical component of magnetic field B_z. Instead, the circular coil is used with the SQUID pick up coil that measures the in plane variation of magnetic flux B_x, figure 4 (right). It is interesting to note that in both the above configurations the SQUID measure only the magnetic field change ΔB.

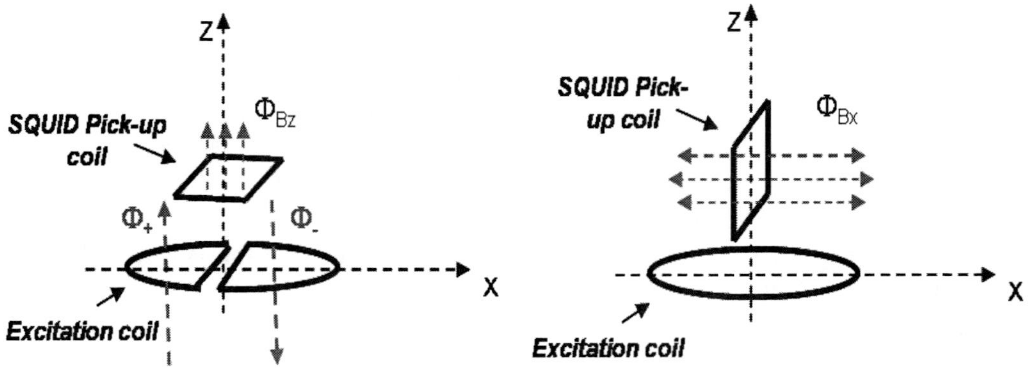

Figure 4. (left) and (right) gradiometeric and circular excitation coil coupled with the SQUID pick up coil to measure the vertical and in-plane component of magnetic field, respectively.

2.3. GMR Sensors for ECT

A typical GMR sensor consist of two magnetic layers, one is magnetically hard while the other is soft, separated by a thin non-magnetic layer (e.g. copper) with a total thickness of few nanometres. The magnetization of the soft magnetic layer (detector) changes its direction under an externally applied magnetic field with respect to the fixed magnetization of the hard layer. As magnetic field sensor the GMR device could be considered as a transducer because it converts the applied magnetic field into electric signals (e.g. voltage).

Magnetic field sensors based on the GMR-effect are generally used in magnetic field sensing applications, such as medicine and industry fields, since they represent a good compromise between magnetic field sensitivity and simple handling. GMR-sensors offer advantage such as high thermal stability, low power consumption, wide linear range of operation, dc to up 1 MHz frequency response, and a small size, when compared with search coils, flux gates, and much more intrinsically sensitive device (e.g. SQUIDs and spin resonance magnetometers) [15 -16].

GMR sensors are rugged and show high bandwidth and electronic stability, these properties makes them ideal for detecting eddy currents [17-18]. An excellent example of such applications can be found in the Non-Destructive Evaluation (NDE) of metallic structures used in the aerospace industry. Typically, material inhomogeneities, cracks and internal defects are monitored by eddy currents technique. With respect coil probes, GMR sensors offer two advantages: a flat sensitivity over a wider frequency range and smaller sizes, a feature very important for sensors array configurations. An example of user interface system used to drive the GMR eddy current probe is shown in Figure 5.

With the aim to reduce the environmental noise and increase the signal dynamic range of the GMR probe into eddy current applications, a second order electronic gradiometer can be used.

The gradiometer consists of three GMR-sensors located vertically with a baseline of 5 mm. The GMR sensor closer the sample (1st GMR) is a detector while the other two sensors provide as a reference to subtract the environmental noise. The circular excitation coil, with a diameter of 2 mm, is located below the 1st GMR, parallel to the sensitive axis of GMR sensors realizing a zero field detector.

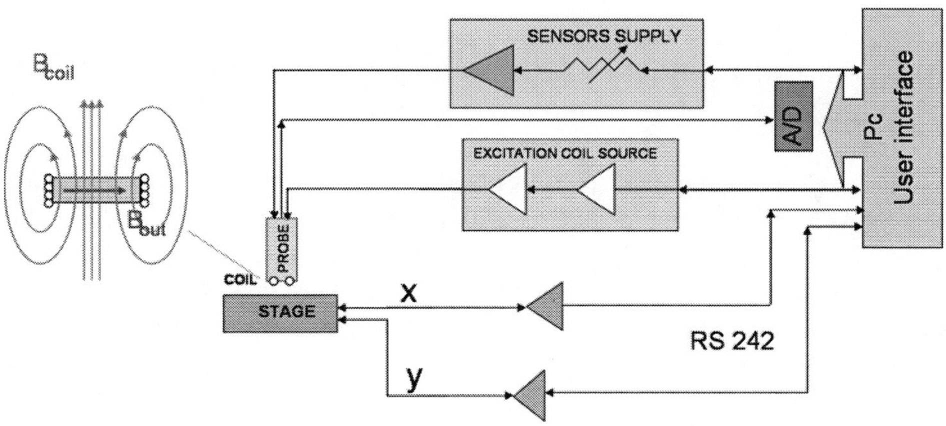

Figure 5. Scheme of the eddy-current GMR-based NDE prototype. The sensor-excitation coil coupling is shown on the left.

2.4. Fluxgate Magnetometer for ECT

The Fluxgate magnetometer is a solid state device used in several applications, such as magnetic airborne detection, search and surveillance operations, non-destructive testing of materials, palaemagnetism and exploration of magnetic field in space missions to the outer planets. For an exhaustive review about the fluxgate applications refer to [19].

This device is a directional magnetometer which working principle is based on the non-linearity of the magnetic characteristic of ferromagnetic core material in its sensing element. It measures the d.c. or low-frequency a.c. component of the magnetic field parallel to the axis of the sensing element, with a sensitivity which ranges between 10^{-11} to 10^{-4} Tesla. In its simplest configuration the sensor consists of a cylindrical core of high permeability material around which two coaxial coils have been wound.

The fluxgate provides a voltage which is proportional to the applied field, the basic analytical description may start from the Faraday law:

$$V_{ind} = d\Phi/dt = d(NA\mu_0\mu(t)H(t))/dt)$$

where $\Phi = BA$ is the magnetic flux in the material and A is the area of the sensor perpendicular to the magnetic field; $H(t)$ is the magnetic field in the sensor core and $\mu(t)$ is the sensor core relative permeability. The permeability of the core is usually modulated by superimposing an a.c. magnetic field which causes the permeability to go trough a maximum twice per drive cycle. This produces a voltage proportional to B_{ext} at the second harmonic of the drive frequency.

A gradiometric probe based on three flux gate sensors has been used to eddy current applications. In particular, the three flux gate are vertically aligned with the sensible axis horizontally oriented so that the in-plane component of the magnetic field has been measured. The flux gate close to the sample is the detector while the other two are used as a reference. An excitation coil with a diameter of 25 mm was coupled below the flux gate detector. The electronics of the system based on the flux-gate gradiometer works at room temperature.

3. Magnetic Field Resolution of Electromagnetic Probes in Eddy Current Aeronautical Applications

Aeronautical applications require techniques capable of detect crack, fatigue and corrosion in Al alloy an Titanium alloy metallic structure. In aircraft industry the quality tests during the fabrication process or integrity assessments of structures already in use need a high accuracy in the localization of both surface and deeply embedded defects (depth >8 mm) in metallic structures [9-11]. These defects can be detected using magnetic probe having a sensitivity ranging between the nano-Tesla to few pico-Tesla working at low frequencies, less than 10 kHz.

Currently, different kinds of electromagnetic probes such as induction coils, SQUID, GMR and Flux-gate, are used for EC-NDE measurements. Typically, the electromagnetic sensors listed are attractive as multimode instruments capable of extremely high magnetic field sensitivity (detection of few pT signals), even in unshielded environments, high image resolution for low frequency eddy-current distributions, large bandwidth (up to MHz) and high spatial resolution [4-5].

Naturally, the need to operate the SQUID at cryogenic temperatures limits its use in many practical applications, but the advent of High Critical Temperature Superconductors (HTc) and the development of high performance HTc SQUIDs has renewed the interest for NDE employing superconductive sensors [6-8].

Since each of the sensors has pros and cons, it is difficult to state which one is the most effective for a particular application and a compromise between the magnetic field resolution and the cost has been required. In Figure 6 the magnetic sensors used in NDT are summarized including their sensitivity and the cost. In the latter figure it is interesting to note that although the GMR sensors are less sensitive than SQUID and search coil, the low cost of the sensors joint to its micrometric dimension highlights good potentiality to develop cheap and rouged probes for eddy current for NDE applications.

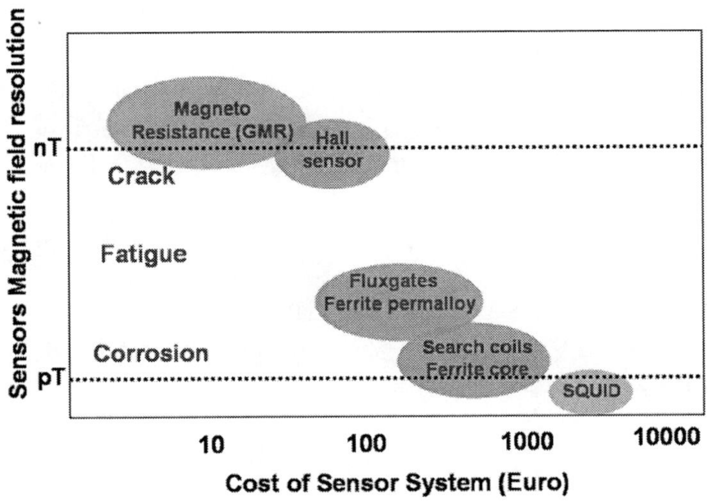

Figure 6. Magnetic and induction coil sensors sensitivity versus the cost. The magnetic intensity of the main defect , for aeronautical applications, in the range between nT and pT are shown.

4. Magnetic Field Noise Resolution of Electromagnetic Sensors

In EC-NDE measurements, noise and signal response depends on the particular defect and it is difficult to measure them separately [11]. Therefore, the magnetic field noise that characterize each sensor must be considered to evaluate correctly the electromagnetic sensors signal output. The magnetic field noise density of the electromagnetic sensors have been compared and reported in Figure 7. An experimental overview of the magnetic field resolution of the electromagnetic sensors can be found in [21].

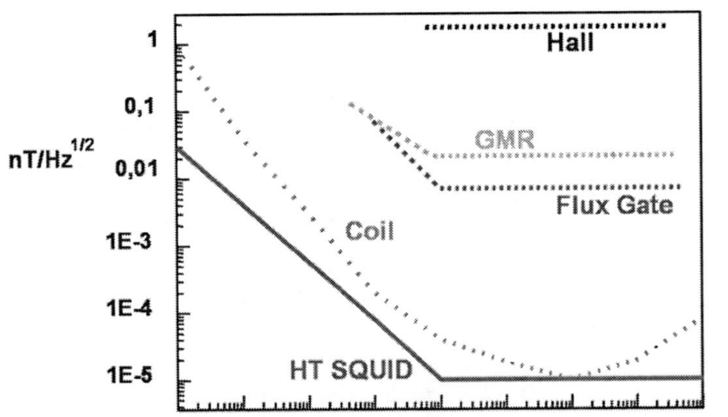

Figure 7. Magnetic field resolution of the magnetic sensors discussed in the text. The magnetic sensors has been compared with the magnetic field resolution of a coil.

From the data shown in the Figure 7 it could be noted that SQUID has the highest magnetic field sensitivity respect with GMR gradiometer, Flux gate and induction coils. Moreover, in unshielded environment i.e. in industrial area, the magnetic field resolution of the SQUID is about 0.3 pT/√Hz at 1 kHz. Moreover, the electromagnetic sensors have several parameters, such as slew rate, dynamic range and the frequency bandwidth, that depend on the frequency.

Usually, the dc SQUID magnetometer is characterized by a slew rate up to 10^3 Φ_0/s (at 1 kHz) a wide bandwidth (from dc to 26 kHz) and a broad dynamic range (130 dB). Otherwise, the gradiometer fabricated with GMR show a spatial resolution better than 60 μm and a 3 dB bandwidth at 700 kHz, while the flux-gate has a voltage noise density of 4.3 μV/√Hz in the frequency range 10 Hz-4 kHz.

4.1 . Lift–off Variation

The analysis of the signal attenuation due to the variation of Lift-Off (LO) between sample and probe, is of great importance to evaluate the performance of the probe used in eddy current testing. In this paragraph, in order to compare the different quantities recorded as a function of the position, the peak-to-peak signal amplitude S_{pp}, detected at distance z above the sample, has been normalized to its maximum value S_{max} (S_{pp}/S_{max}) for magnetic and inductive probes. In Figure 8 experimental data show a sharper decrease (up to 90 %) of the induction coil signal respect to the magnetic probes in correspondence of LO values ≥ 5 mm.

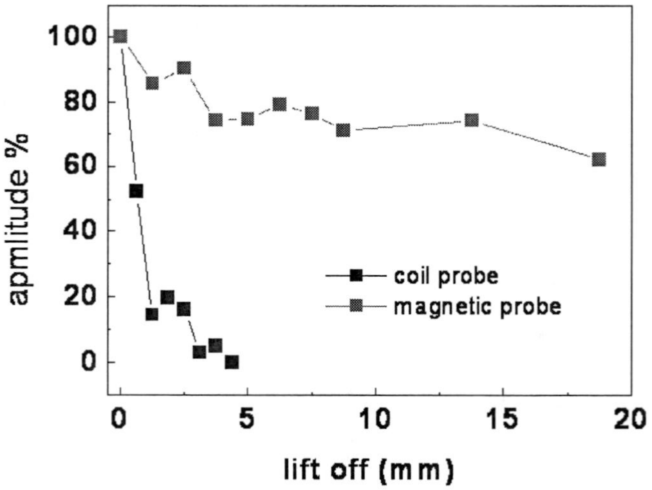

Figure 8. Peak-to peak signal amplitude Spp normalized to its maximum value Smax as a function of lift-off between probe and sample. Black square are relative to coil probe while red square to magnetic probe, respectively.

For the magnetic sensors the same signal attenuation is reached for a LO ≥ 15 mm. In the case of weak signals, generated by small or deep flaws, the strong dependence of induction coils from the lift-off variation reduces the accuracy and the reproducibility of the measurements. Moreover, the strong signal attenuation, due to high lift-off values, limits the use of these probes in applications where, for example, the conductive layer under examination is covered by a thick electrically insulating layer, like in composite structure largely used in aeronautical industry.

4.2. Probe Tilting

Another possible source of noise in EC-NDE measurements is due to the accidental tilting of the probe axis with respect to the direction of the sample plane. In this case the contribution of other components of magnetic field can change the output signal. In the follow the experimental data collected using both magnetic and conductive probes has been compared. The probes have been tilted with an angle ranging from 0° to 6.5°. The axial magnetic field component B_z as a function of the line-scan coordinate has been recorded. The relevant quantities, such as the peak-to-peak amplitude S_{pp} as a function of the tilt angle, are shown in Figure 9. The signal amplitude variation due to a tilting of 1° for the SQUID probe is 76 pT. It is hard to perform this experiment for the massive induction coils where a tilt angle of 1° already nullifies the amplitude of the detected signal!

Studies of the effect of a probe tilt angle reveal that the magnetic probes signal amplitude decreases less than coils. For this reason the magnetic probes can be considered, to a very good extent, as a point sensor, which is simplifying both the measurements and the numerical simulations.

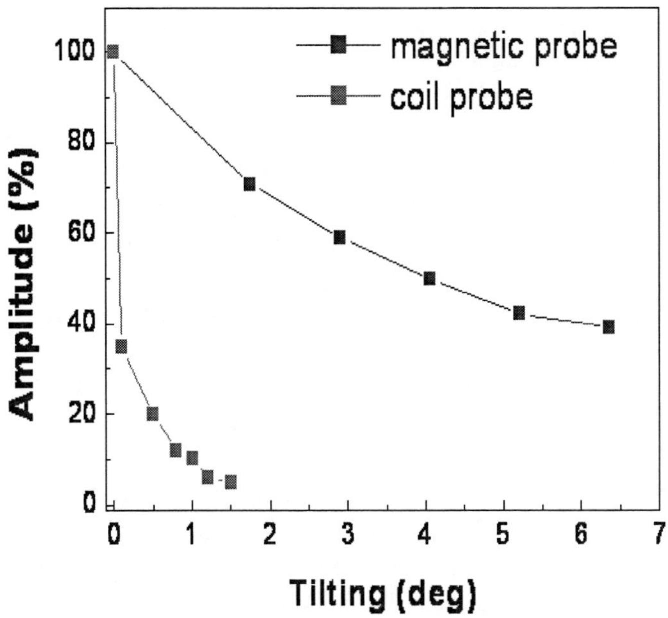

Figure 9 - Signal amplitude value of the detected axial component of the magnetic field Bz and the impedance variation as a function of the tilt angle between the probe and the plane. The red curve is relative to coil while the blue to the magnetic probe.

5. Detection of Deep Flaws in Metallic Multilayer Structures

With the aim to perform a comparison between the different eddy current sensors, an independent parameter has been chosen. It is a type of Signal-to-Noise ratio (S/N) defined as the ratio between the maximum signal variation along the measured line-scan (S_{pp}) and the standard deviation of the N data points b_i collected along the same line scan, i.e. $n = [\sum_i (b_i - b_m)^2/N]^{1/2}$, where b_m is the average value of N points [22]. Since in our measurements line-scans are crossing a consistent area of the sample located far away the defect, n can be considered in each measurement as the noise. To compare the probe sensitivity to deep flaws, output signals, as a function of the line-scan coordinate, were recorded.

Several Aluminium alloy samples (Al-T4) has been used in this experiments. They ware designed to test the sensitivity of the sensors in different conditions. Several tests on single-layer structures have been performed with the aim to gives information on the detection of multiple slots at depths greater than 10 mm. Moreover, three-layer structures have been tested to give information on the detection of a long at even larger depths (20 mm and 40 mm), while 5-layers structures have been analysed to detect millimetre-sized defect at depths larger than 10 mm. The presence of many layers and insulating gaps between them results both in a significant reduction of the conductivity of the whole structure and in a complex wave fields reflection at the interfaces. As a consequence the attenuation of the output signal occurs.

In Figure 10 the comparison of the sensitivity of the three probes to slots 11 mm and 12 mm deep is shown. The excitation coil coupled with the magnetic probes, SQUID and Flux-gate, was a gradiometric coil [23] with a diameter of 25mm, working at an operating frequency of 77 Hz, and biased with a current amplitude of 120 mA and 180 mA, respectively. Instead, the absolute inductive coil used in the test had a frequency of 70 Hz. and a diameter of 38mm. The S/N ratios estimated for the 12 mm and 11 mm deep slots were 4.20 and 1.95 for the SQUID, 2.46 and 0.64 for the flux-gate, 0.65 and 0.25 for the induction coil. This result, confirmed by the dependence of the S/N ratio on the frequency (Figure 11), highlights the high sensitivity of the SQUID when compared to the other probes.

The detection of the very deep crack (20 mm and 40mm), not reported here, was performed by the SQUID with a gradiometric coil fed with 180mA at 13 Hz. The relative values of the signal-to-noise ratio is S/N = 2.78, while the induction coil did not give measurable signal for this test.

The experimental results related to the test on multilayer assembled with 5 single-layers of Al T4 alloy, with insulating gaps between the stacked layers, are summarized in Figure 11.

Figure 12 (a) shows the experimental results obtained with the flux-gate at 133 Hz and 77 Hz using the gradiometric coil with a current of 180 mA. The output signal in this case did not reveal the defect even if the frequency is reduced. The flux gate probe was able to detect the slot signal with a S/N = 1.85 only removing the upper layer, (Figure 12 (b)). Instead, the induction coil, operating at 70 Hz, detected the hole with a S/N = 0.16. The signal amplitude output improve only removing the third and second upper layers, as shown in Figure 12 (c). Finally, Figure 12 (d) shows the SQUID response for three different frequencies (133 Hz, 177 Hz and 577 Hz) using a current of 120 mA. The best signal-to-noise ratio S/N = 3.573 is obtained at 133 Hz, while it decreases slightly at 177 Hz while at 577 Hz any detectable signal was observed. The results depicted in Figure 12 demonstrate clearly the superior performance of SQUID in the detection of deep defect.

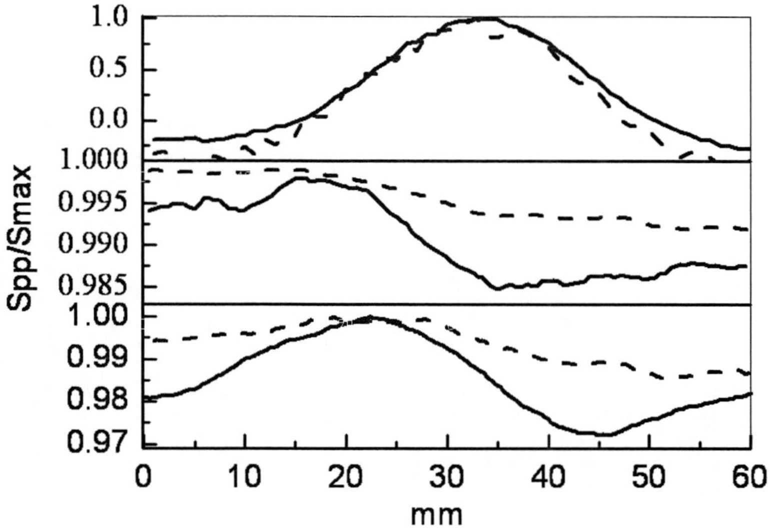

Figure 10. Peak-to-peak signal amplitude Spp normalized to its maximum value Smax as a function of the line-scan coordinate. Measurements of the 12 mm deep (dashed line) and 11 mm deep (full line) slots are shown for induction coil (above), flux-gate (middle) and SQUID (below).

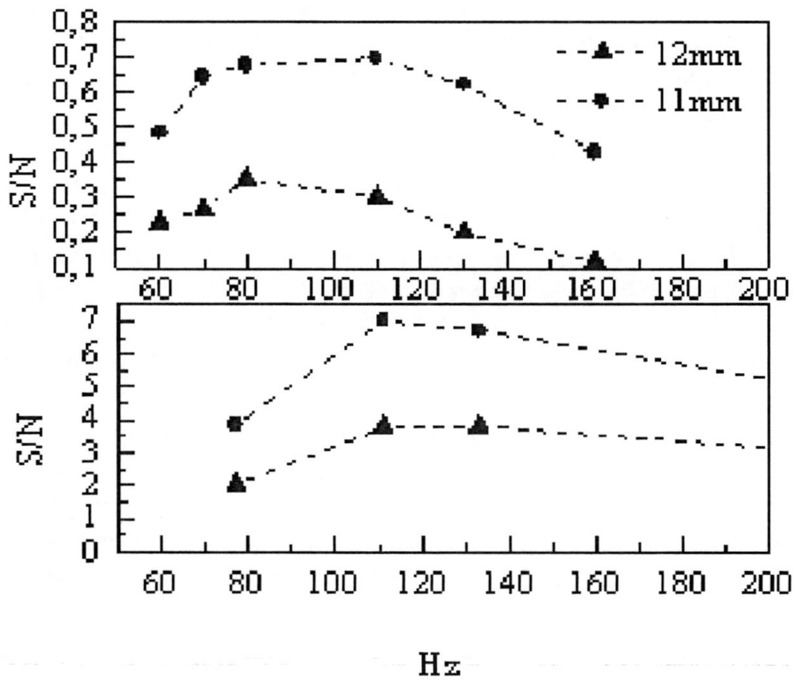

Figure 11. Dependence on frequency of the signal-to-noise ratio (S/N) of the induction coil (above) and SQUID (below). Triangles, and circles are relative to the 12 mm and 11 mm deep slot, respectively.

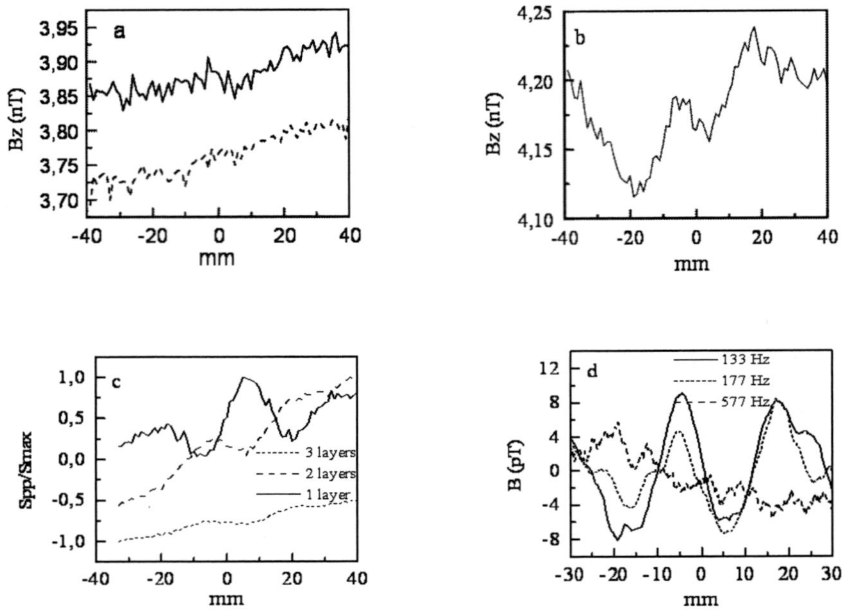

Figure 12. Experimental results for the flux-gate, with the gradiometric excitation coil fed with 180 mA, on a 5 layer structures (a) and removing the upper layer (at 133 Hz), (b) ; induction coil (c) at 70 Hz in 3-layers, 2-layers and monolayer structure, (d) SQUID response in correspondence of three frequencies (133 Hz, 177 Hz, and 577 Hz) using 5 layers sample.

It could be noted that the SQUID as EC probe, is the only sensor that detects successfully the defects in multilayers structures where the presence of voids and insulating layers, produce a reduction of the conductivity and complex reflections of wavefield at the interface. As demonstrated, the SQUID is the only magnetic sensor capable of detecting these anomalies with sufficient sensitivity.

The results relative to a quantitative comparison between different eddy-current probes (magnetic and induction coil) has been collected in Figure 13. It is interesting to note that the data reported in the previus figure are in good agreement with the skin depth law, represented by the continuus line. For this reason it is not posible to distinguish the magnetic probes data from the data measured using inductive probe.

In the Figure 14 the signal-to-noise ratios related to the many experiments performed at different depth in metallic sample are reported as function of crack depth and frequency. It is interesting to note that both magnetic and coil probes are in good agreement with the electromagnetic skin depth law (1) reported above for sub-superficial and very deep defects.

Finally, the results reported in this paragraph show the performances of the different probes in their optimized operation conditions. The signal-to-noise ratio (S/N) showed in Figure 14 demonstrates the superior performance of the magnetic probes, in particular SQUID, in detecting subsurface and deep defects at depths greater than 20 mm respect to the coil.

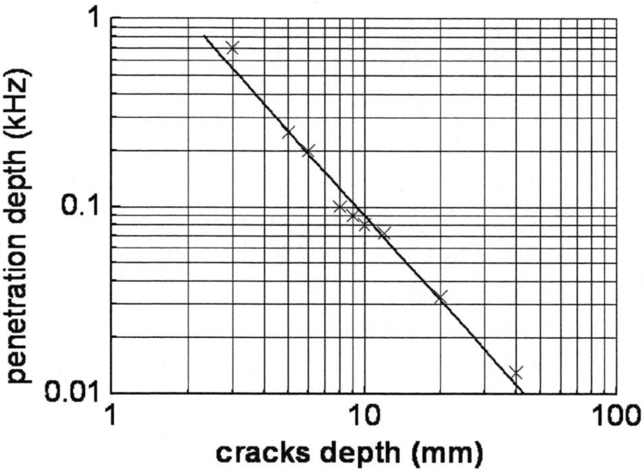

Figure 13. Detection capabilities of different eddy-current probes using magnetic and induction coil sensors (x) versus the depth of the defect. The skin depth law is represented by the continuous line and is calculated using the equation (1).

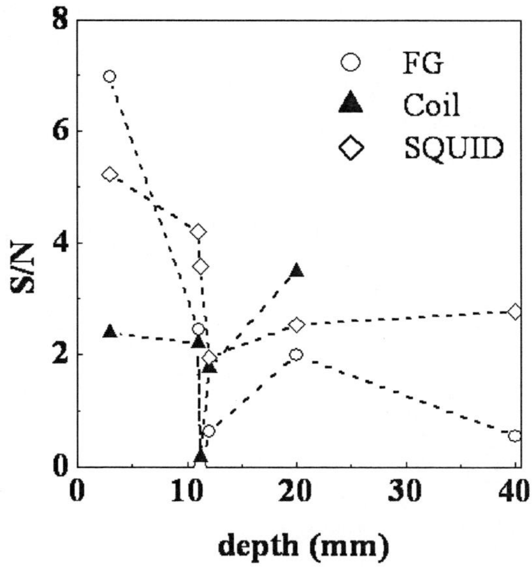

Figure 14. Signal-to-noise ratio S/N versus the defects depth.

6. Detection of Defect in Fiber Metallic Laminate Material

Since FMLs combine the best features of fibre reinforced composites and metals, it is possible to overcome some drawbacks of the base materials employed. Usually, FMLs consist of alternating thin sheets of high strength aluminium bonded to fibre-epoxy layers [24]. Since FMLs are employed principally in aircraft fuselage structures, it is important to investigate their damage features such as debonding and porosity due to manufacturing process. Usually, these defects are not visible and an high sensitivity inspection techniques are required. Our aim is to show the capability of advanced magnetic sensors like SQUID and GMR to detect the presence of debonding and porosity inside composite materials.

FML tested in this work was made of 2024 T3 aluminium 0.3-mm thick sheets and FM94 S2-glass/epoxy pre-peg 0.125-mm thick layers. The resin content in the pre-peg was 27% by weight. The stacking sequence of the specimens is shown in Figure 15. The artificial defects included in the test sample are pieces of Teflon called *A* and *B*, with a dimensions of 3x3 mm and 6x6 mm, respectively. The pieces *A* are located under the first aluminium layer, above the glass layers, while the pieces *B* are included above the second aluminium layer under the fiber glass layers.

To simulate a porosity inside that specimen, a piece of wet cotton has been introduced between the glass layers and the second aluminium plate. A sketch of the artificial defect into the Glare layers are reported in Figure 16.

The specimen was inspected by NDE tests by using eddy current technique based on SQUID and GMRs gradiometer equipments developed at the laboratories of the CNR INFM Coherentia (Napoli-Italy). More details about the characteristics of the SQUID sensors used in the inspection of FML can be found in the references [25-26]. The SQUID used to test the specimens is a planar gradiometer that has a field sensitivity in unshielded environment less than 1 pT-rms for frequencies above 100 Hz, a slew-rate of 10^3 Φ_0/s and a dynamical range of about 130 dB. The SQUID gradiometer is oriented so that the z-axis component of the magnetic field is measured. The eddy current source is a wire-wound circular coil with a diameter of 5 mm located under the dewar at 2-mm far from the specimens. The stand-off is about 6 mm. The feed current was 10 mA-rms at the frequency of 22 kHz. In the case of GMR gradiometer the measurement was carried out using a coil of 2 mm fed with a current of 135 mA at 35 kHz.

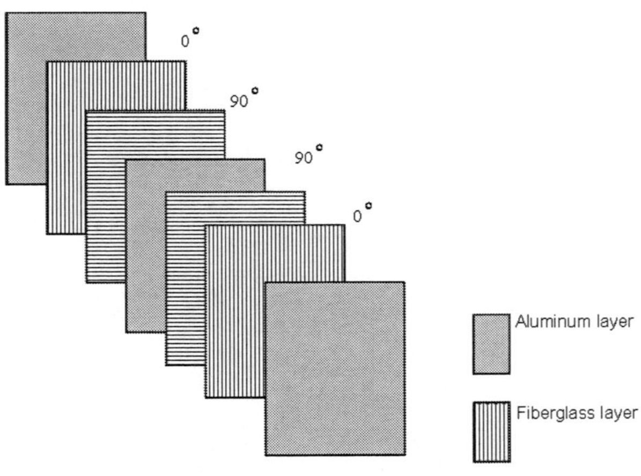

Figure 15. Stacking sequence of the FML specimens inspected. The orientation of the fibre glass is also reported.

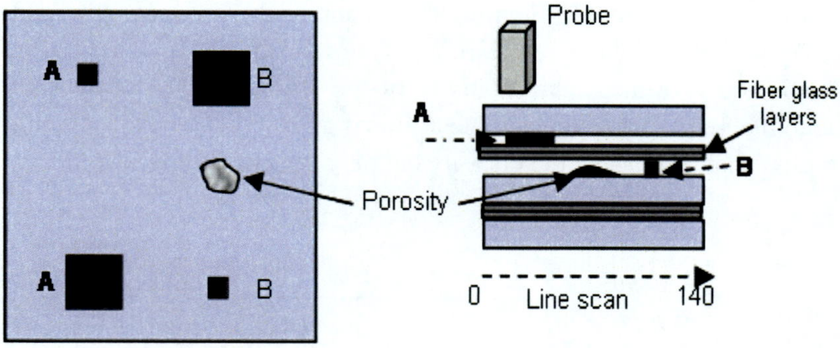

Figure 16. The scheme of the artificial defect included into the test specimen.

In Figure 17 (left and right), the magnitude of the magnetic field, measured by using the GMR and SQUID gradiometer on the sample area that includes defect *A* (3x3mm) and *B* (6x6mm), is shown. The line scan reported in Figure 17 (left) represents the in-plane x-component of the magnetic field. The output represented in Figure 17 (right) is the z-component of the magnetic field. It could be noted that the peak-to-peak signal detected by both sensors in correspondence of the two inclusions is different. In the *A* case, concerning the piece of Teflon, the amplitude of the signal is lower than the signal related to the defect indicate as *B*. The two eddy current equipments detect successfully the Teflon inclusions, they are capable to localize the position of the defects and recognize the different dimensions and depth of the defects.

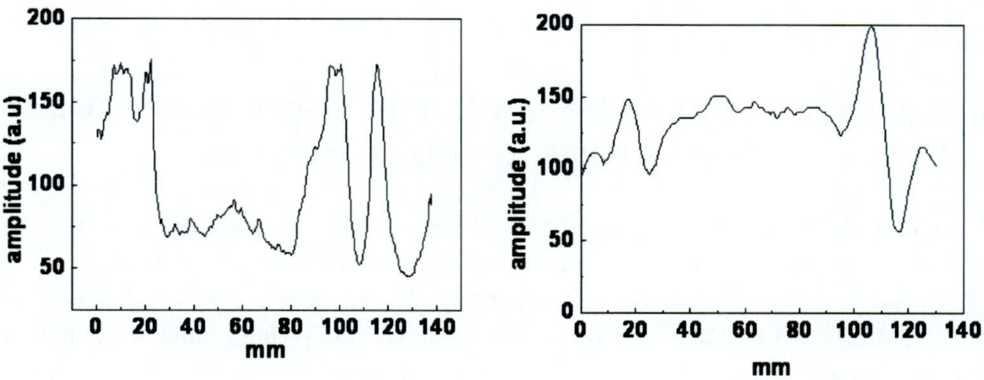

Figure 17. (left) and (right) the lines scan representing the normalized output signals obtained by using GMR and HTs SQUID gradiometers, respectively. On the line-scans the defects A (3x3mm) and B (6x6 mm) appear from left to right, respectively.

Moreover, one of the open challenge concerning the NDE investigation of composite multilayers materials is the detection of porosity inside them. In the test sample presented here a wet cotton was included to simulate this kind of damage and to test the eddy current equipment based on GMR and HTS SQUID gradiometer. In Figure 18 the magnetic field imaging of the specimen area that includes the artificial porosity between two Teflon pieces

(6x6mm and 3x3mm) is reported. In Figure 18 (left) and (right) the magnetic field output of the HTS SQUID and GMR gradiometer, respectively are shown.

It is interesting to note the high sensitivity of the HTS SQUID to detect the artificial debonding due to the Teflon inclusion and the artificial porosity. On the other hand, the GMR gradiometer is not capable to detect the porosity but only the inclusions.

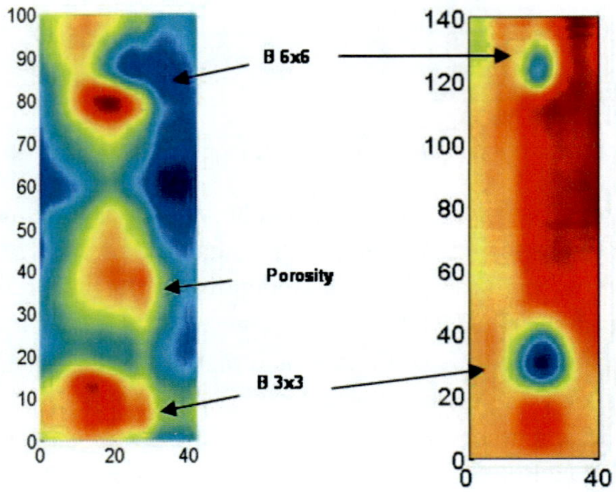

Figure 18. Eddy Current imaging of the specimen in correspondence of the artificial porosity between two Teflon inclusions. On the left and right the signal output of HTS SQUID and GMR gradiometer, respectively are shown.

7. Characterization of Defect Due to Low Velocity Impact in Carbon Fiber Reinforced Plastic

Composite materials, specifically Carbon Fibre Reinforced Polymers (CFRPs), offer excellent mechanical behaviour, good damage tolerance, high strength and rigidity with low weight, good corrosion resistance and suitability for the production of complex-shape components with reduced manufacture time. However, due to low inter-laminar strength fibre composites are susceptible to delamination during processing or in service.

Delamination may be produced, for example in aeronautical applications, by runway debris, hail, maintenance damage (i.e., dropped tools) and bird strikes. Frequently, delaminations between the layers develop in fibre fractures with no visible surface manifestation. CFRPs are capable of absorbing the energy of impact thanks to the presence of a polymeric matrix that distributes the energy in the material. In this way the impact doesn't produce a perforation of the composite structure but may cause internal damage such as delamination. The presence and the growth of delamination may produce severe stiffness reduction in the structure, leading to a catastrophic failure. The degree of damage depends on various factors, for example, the energy of the impact, the thickness of structures and the fibre orientation of layers in the composite.

A range of experimental results over the last decade has shown that the mechanical deformation and the electrical resistance of CFRPs are coupled. Both the mechanical properties and the level of damage of CFRPs are correlated to the electrical conductivity of the reinforcing carbon fibres. It was demonstrated that the variation of fibre distance or fibre breakage changes the resistivity of the reinforced composite [27,28]. In this way, delaminations and fibre breakage alter the electrical conductivity of the material produced by the variation of the current distribution in the composite. As a consequence a variation of the corresponding magnetic signal produced by the eddy current into the sample occurs. The low electrical conductivity and high anisotropy of CFRP composites generally doesn't allow conventional techniques, such as traditional ultrasound, thermography and eddy current with induction coils, to detect early stage damage in composite samples. In fact the thermography was found unreliable when it tests bonded joints with a narrow gap between the unbonded surface, or when the damage in the material is few millimetres in depth [29]. On the other hand, the ultrasound technique isn't very sensitive when the material has a rough surface and a layered structure [30,31]. Moreover, the eddy current technique shows a reduction of the sensitivity of induction coil to evaluate deep flaws in structures characterized by a low electrical conductivity [32].

For this reason an eddy current method utilizing an HTS SQUID magnetometer, thanks to its high magnetic field sensitivity ($\cong$100 $fT/Hz^{1/2}$ in unshielded environment), could be an effective technique to improve the non-destructive evaluation of CFRP materials.

The CFRP samples shown in Figure 19 are multidirectional composites 70 mm x 70 mm based on an epoxy-matrix (HMF 934) reinforced with prepeg[1] layers made of T400 carbon fibres fabricated by hand lay-up[2] and autoclave curing. The fibre content in the composite was 55% by volume. Each sample is made of six blocks of layers stacked together, and each block has a number of overlapping layers with the same fibre orientation, as shown in Figure19. The number of layers in each block depends on the total thickness of the sample, which ranges from 2 mm to 4 mm.

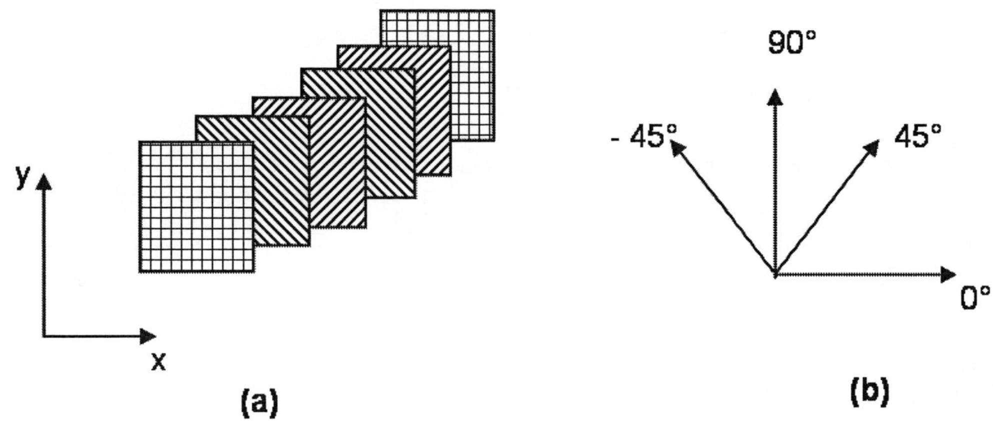

Figure 19. a) Stacking sequence of a multi-layer sample [(0/90),±45]S; b) fibre orientations.

1 A reinforced-plastics term for the reinforcing material that contains or is combined with the full complement of resin before the molding operation.
2 Production of reinforced plastics by positioning the reinforcing material, such as glass fabric, in the mold prior to impregnation with resin.

The multidirectional specimens were loaded at the centre by a hemispherical steel indenter, having a diameter of about 13 mm, and falling from different heights using a drop weight tower machine. Specimens with a thickness of 2-mm were impacted with an energy level up to 12 J, while 3 mm and 4 mm thick samples were damaged with impact energies up to 25 J. In particular, we focus our attention on 2-mm thick samples impacted at 2.4 J, 7.8 J, 12 J, and on 4 mm thick sample at 1.8 J. The variable depth of the damage, i.e. indentation, located at the centre of the sample was measured using a micrometer.

In Figure 20 (a) and (b) the magnitude and the phase of the magnetic field for the virgin (α lines) and the damaged (β and γ line) 2-mm-thick samples are presented. The β and γ line-scans were obtained moving the sensor across the center of the defects due to an impact energy level of 2.4 J and 12 J respectively. Each line represents a 60 mm scan over the sample. The magnetic field magnitude for the damaged samples (β and γ lines), in the vicinity of the defect center (x = 0), shows no significant variation. For this reason, their change in magnitude with distance looks quite similar to that for the virgin sample. Apart from that, the only difference that could be noted among of the line-scans is the different value of the magnetic field amplitude. The magnitudes of the damaged samples are lower than the magnitude of the virgin sample; in particular, the higher the damage, the lower the magnitude of the magnetic signal. So, considering only the magnitude of the magnetic field, the presence of the defect is not obvious.

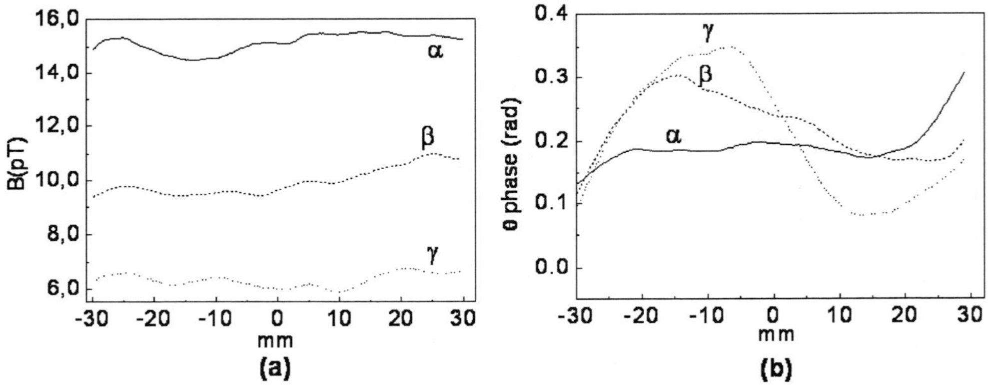

Figure 20. (a) and (b) represent the magnitude and the phase of the magnetic field for 2mm thick specimens. (α) is the line-scan related to the virgin sample, (β) and (γ) are the line-scans of the damaged samples impacted at 2.4 J and 12 J respectively.

Instead, the phase of the magnetic field related to the damaged samples presents a variation along the line-scan. The difference between the signal of virgin and damaged specimen could be predicted by the slope of the phase signal among the x = 0 coordinate. An inflexion that corresponds to a presence of damage characterizes the lines of impacted samples.

Moreover, for specimens with a visible damage the variation of the magnetic phase is correlated with the damage location. In particular, we have checked that the defect center coincides with the center of the inflexion (x = 0).

It should be noted that the decreasing and increasing phase value, at the beginning and the end of the scan is purely edge effects. This is due to the comparable dimension of the specimen and the line-scan length. Experimental results demonstrate that the slope of the magnetic phase (*dθ/dx*) is correlated with the damage severity. In other words it is possible to discriminate the different damage using the slope value, in fact the greater the slope value of the magnetic phase the more visible the damage and the greater the defect depth [34].

In Figure 21 (a) the trend of the slope value against the impact energy levels is shown. For each thickness increasing the impact energy increases the slope value. In the case of 2 mm thick samples there aren't slope values for energy higher than 12 J because the specimens are already perforated, as is evident from an inspection of the sample. Comparing the slopes for each impact energy, the values for the 2 mm thick samples are always higher than those for the 3-mm and 4-mm thick samples. It is reasonable that specimens with small thickness are more brittle than samples with a bigger thickness, for this reason also comparable impact energies produce higher indentations in the 2 mm thick specimen. The slope is similar at different value of indentations in the case of the 3 mm and 4 mm thick samples. In other words composites with smaller thickness have a lower mechanical tolerance than specimens with bigger thickness. It should be noted that the experimental results shown in Figure 3(a) do not depend on the particular frequency chosen for the excitation coil. As shown in Figure 3(b) the slope of the magnetic phase measured by the NDE SQUID based system follows a linear behaviour versus the frequency of the excitation coil in the range 5 kHz-25 kHz.

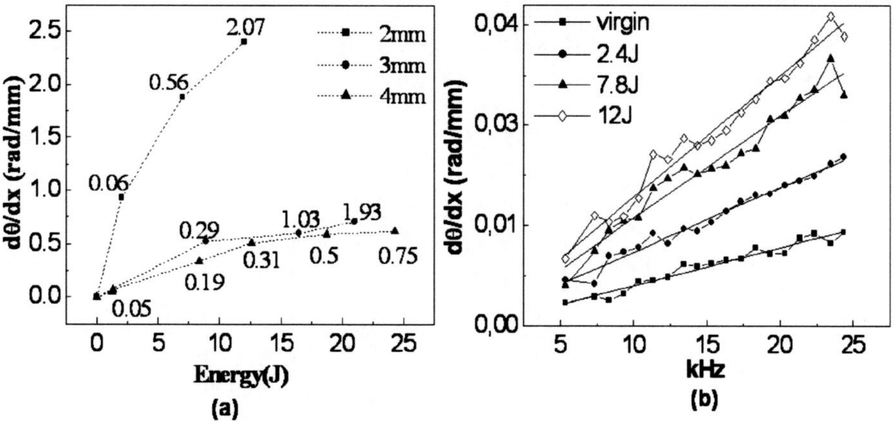

Figure 21: (a) slope of the phase of the magnetic field versus the impact energy level for 2 mm, 3 mm and 4 mm thick specimens. In correspondence of each slope value the defect depth or indentation (in mm) is reported; (b) slope of the phase of the magnetic field versus frequency of the excitation coil for the 2 mm thick samples impacted at different energy levels. The error bars related to the slope value are less than 2%.

Previous mechanical tests and optical analyses of the cross section of the damaged sample area, demonstrated that 4 mm thick samples for energy higher than about 1.6 J are susceptible to early internal delamination. For this reason in the following investigation, we have considered an impact energy level of about 1.8 J for a 4 mm thick specimen.

In Figure 22(a) and 22(b) we show the image of magnetic phase for a 4 mm thick virgin and impacted sample with an energy level of 1.8 J respectively. The image of the damaged sample (Figure 22(b)) is obtained without any data processing. The dipole-like image is related to the non-homogeneous eddy-current distribution around the center of the impact located at the position (0,0).

It must be noted that in the picture of the damaged sample Figure 22(b) the edge effect, which characterizes the virgin map in Figure 21(a), is negligible compared with variations of the phase of magnetic field due to the damage inside the specimen.

Though the impact energy is very weak, the magnetic signal reveals the presence of the defect even when it isn't visible to naked eye. It appears more evident in Figure 23, where we compare the line-scan (taken from the maps of Figure 22) at the center of impact point (line (d)), at ±20 mm from it (lines (b) and (c)), and the line-scan of the virgin sample with the same thickness (line (a)). The slope variation between the lines (a) and (d) means that, although the sample surface appears to be intact, the sample has internal damage such as delamination.

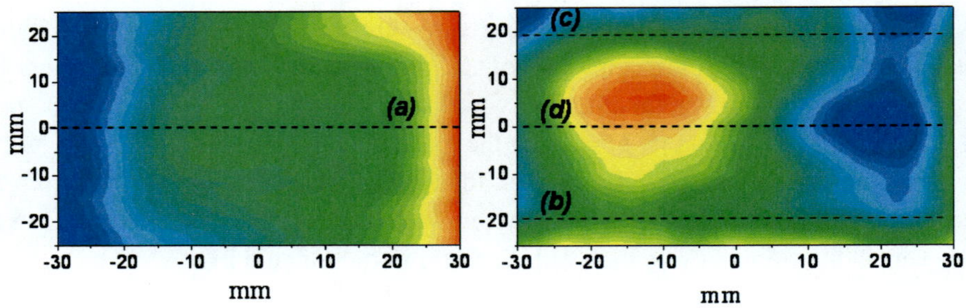

Figure 22: Image of the magnetic field phase (rad) for cross-ply composite: (left) 4 mm thick virgin sample; (right) magnetic image of 4 mm thick sample impacted with an energy level of 1.8 J.

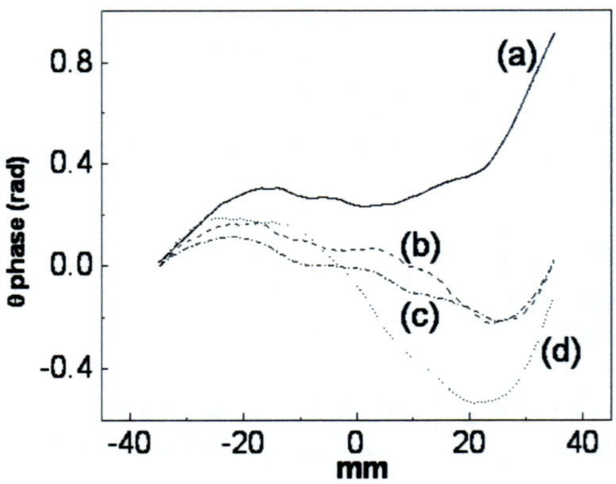

Figure 23: line-scan taken from the map in the Figure 22 (left and right). In the latter (a) is the line of virgin sample 4 mm thick, (b) and (c) are the line-scans located at ±20 mm from the center of impact point, (d) is the line-scan at impacted point.

Moreover, we focus our attention on the comparison among the virgin line-scan and the line-scans collected at ±20 mm from the center of the impacted area. Even though the 4 mm thick sample is damaged with a very low energy level, U =1.8 J, the damage propagates 20-mm away the center, as shown from the non negligible slope variation of the magnetic phase with respect to the virgin one. It means that the variation of the magnetic phase along the line-scan expressed by the slope $d\theta/dx$ is a suitable parameter to investigate at an early stage not only the severity of the damage but also its extension inside the laminates. It is worth noting that the detection of damage due to very low values of impact energy considered in this work generally fails using conventional eddy current techniques with induction coils.

The maps shown in Figure 24, that report the image for the same samples considered in the Figure 22, demonstrate this statement. The eddy current analysis shown in Figure 24 was performed using an Elotest B300 instrument supplied by Rohmann GmbH. The maps were obtained using an absolute ferrite core probe with an excitation frequency of 580 kHz. It should be noted that the conventional eddy current technique based on the induction coils is not able to distinguish the virgin sample from the damaged one at 1.8 J. Instead, the SQUID ECT probe detect the defect although its dimension are not visible to the necked eye as shown in the Figure 25 (left). This result highlights the difficulties of the conventional eddy current inspection to detect impact at low energy levels (lower than 2 J) [37-39]. In other words the comparison between the images in Figure 24 and Figure 25 emphasizes the high sensitivity of the NDE system based on HTS SQUID magnetometer to detect early stage damage in complex structure like CFRPs.

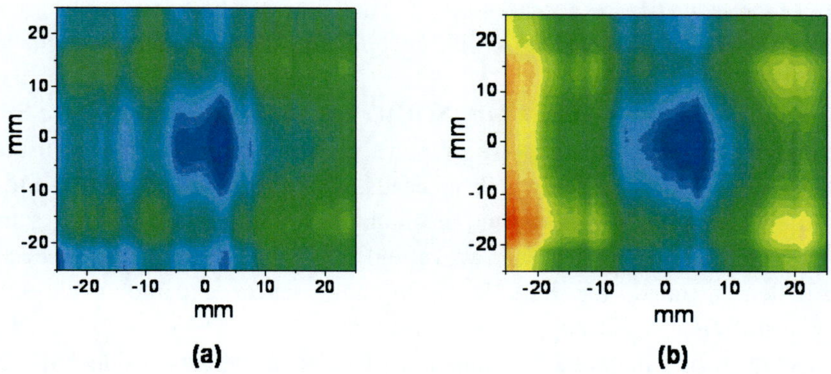

Figure 24: SQUID eddy current image for (a) 4 mm thick virgin sample and (b) 4 mm thick sample impacted with an energy level of 1.8 J.

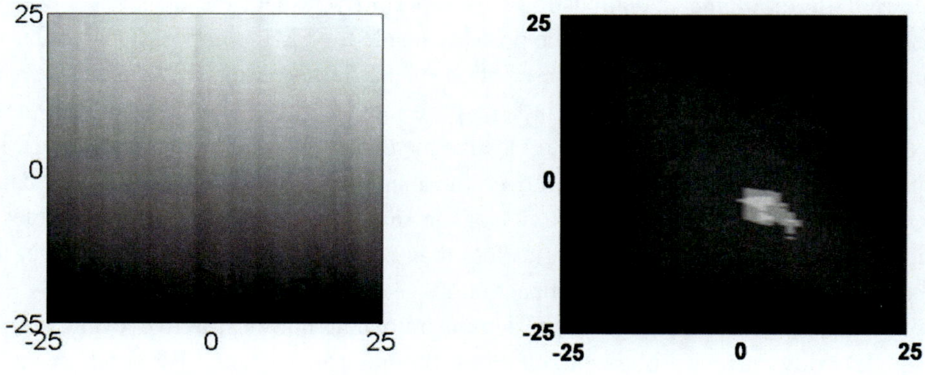

Figure 25: eddy current image using conventional coil probe (left) for 4 mm thick virgin sample and magnetic image SQUID ECT probe (right) on 4 mm thick sample impacted with an energy level of 1.8 J.

Finally, HTS dc SQUID eddy current probe is capable of detecting impact defects at an early stage, immediately following an elastic response, i.e., impact energy less than 2 J in 4 mm thick CFRP composites, in an unshielded environment. SQUID ECT probe successfully localizes the impact area and internal defects, even when they are not visible to the naked eye.

CONCLUSION

In the Section 5 an experimental characterization based on a quantitative comparison between different eddy-current probes has been reported. In particular, the performance of an induction coil, a flux-gate a HTc-SQUID and a GMR sensors has been investigated on the basis of comparative experiments on Al-alloy test samples with artificial defects such as holes, slots and cracks.

The effect of larger probe lift-off distances is less negative in the case of the magnetic probes than coil. Studies of the effect of a probe tilt angle reveal that the magnetic probes can be considered more sensitive respect to the coil. The latter can be considered to a very good extent, as a point sensor, which is simplifying both the measurements and the numerical simulations.

Moreover the overall superiority of the SQUID can be appreciated from the S/N ratio in each measurements. It is always equally or more sensitive than the other sensors in each individual case, it is the only sensor which succeeds in identifying the defects in all structures examined. In the presence of voids and insulating layers in multi-layered samples, the conductivity is significantly reduced and wavefield reflections at the interface become quite complex. As demonstrated, the SQUID is the only sensor capable of detecting these anomalies with sufficient sensitivity.

In the Section 6 non destructive evaluation of artificial defects inside FML composite materials by using an HTS SQUID and GMR gradiometer has been reported. In this it has been demonstrated that SQUID gradiometer detect internal defects that simulate debonding inside multi-layers metallic alloy. This result is obtained even if an eddy current based on GMR gradiometer is used.

The SQUID gradiometer demonstrates its superiority respect the magnetic field sensor, such as GMRs, to detect successfully the porosity inside the FML composite materials.

Finally, in the Section 7 a large description of the detection of early stage defect inside Carbon Fiber Reinforced Plastic used in aerospace industry has been showed. The magnetic image of a damaged sample allows us to localize the defect and the impacted area even if they are not visible to the naked eye. Moreover the magnitude of the magnetic field can give information about the average conductivity of the sample. The former method enables us to investigate damage due to impact energies less than 2 J and the propagation of matrix micro-failure at least 20 mm away from the impact point.

Above all the experimental results demonstrate that non-destructive evaluation using HTS-SQUID eddy current probe is a suitable technique to analyse CFRP structures affected by early stage damage. The detection of structural defects, not visible to the naked eye and not detectable by conventional coil probe is very important for the research and development of these materials. These results suggest also interesting solution for the NDT in aerospace industry at a maintenance level to improve the flight safety of the aircraft components.

REFERENCES

[1] Crane, R.L.; Rummel, W.D. *Non-destructive Testing Handbook 2nd ed.*, McMaster R.C.; McIntire P.; Mester M.L. The American Society for Non-destructive Testing, Columbux, OH, 1986.

[2] Wenk, S.A.,; McMaster, R.C. *Choosing NDT: Applications, Costs and Benefits of Non-destructive Testing in Your Quality Assurance Program,* Columbus, OH: American Society for Non-destructive Testing, 1987.

[3] Mix, P.E. *Introduction to Nondestructive Testing: a training guide,* John Wiley and Sons, Inc., Hoboken, New Jersey, 2005.

[4] Bonavolontà, C.; Valentino, M.; Marrocco, N.; Pepe, G.P. *IEEE Transactions on Applied Superconductivity* 2009, *19*, 808 – 811.

[5] Valentino, M.; Bonavolontà, C.; Marrocco, N.; Peluso, G.; Pepe, G.P. *Sensors Letters*, 2009, *7*. 1-4.

[6] Cochran, A.; Morgan, L.N.C.; Bowman, R.M.; Kirk, K.J.; Donaldson, G.B. *IEEE Trans Appl Superconductivity* 1993, *3*, 1926-1929.

[7] Tarvin, Y.; Zhang, Y.; Wolf, W.; Braginski, A. *Supercond Sci Technol* 1994, *7*, 265-268.

[8] Mück, M.; Kreutzbruck, M.V.; Baby, U.; Tröll, J.; Heiden, C. *Physica C* 1997, *282-287,* 407-410.

[9] Jenks, W.G.; Sadeghi, S.S.H.; Wikswo, Jr J.P. *J Phys D Appl Phys* 1997, *30*, 293-323.

[10] Ruosi, A.; Pepe, G.; Peluso, G.; Valentino, M.; Monebhurrun, V. *IEEE Trans. Appl. Superconductivity* 1998, *9*, 3499-3502.

[11] Valentino, M. et.al. in *Studies in Applied Electromagnetic and Mechanics*, D.Lesselier IOS Press, 1999, 159-170.

[12] Barone, A.; Paternò, G. *Physics and Applications of the Josephson Effect*, Wiley, New York, NY, 1982.

[13] Koelle, D.; Kleiner, R.; Ludwig, F.; Dantsker, E.; Clarke, J. *Reviews of Modern Physics* 1999, *71,* 631-686

[14] Kreutzbruck, M.V.; Allweins, K.; Gierelt, G.; Krause, H.J.; Gartner, S.; Wolf, W. *Physica C* 2002, *368*, 85-90.

[15] Ripka, P. *Magnetic Sensors and Magnetometers*, Arthec House, Boston London, 2001.

[16] Mahdi, A.E.; Panina, L.; Mapps D. *Sensors and Actuators A* 2003, *105*, 271-285.

[17] Dogaru, T.; Smith, S.T. *IEEE Trans. Magn* 2001, *37*, 3831-3838.

[18] Smith, C.H.; Schneider, R.W.; Dogaru, T.; Smith, S.T. in Rev Prog QNDE, American Institute of Physics, Melville, NY, 2003, *22*, 419-426.

[19] Ripka, P. *Sensors and Actuators A* 1992, *33*, 129-141.

[20] Bartington Instrument Inc., 087XGE Oxford England. Low-noise three-axis flux-gate MAG-3.

[21] Robbes, D.; Dolabdjian, C.; Saez, S.; Monfort, Y.; Kaiser, G.; Ciureanu, P. *IEEE Trans on Applied Superconduc* 2001, *2*, 629-634.

[22] Pávó J. et al. in *Electromagnetic Nondestructive Evaluation (II)*, Albanese R.; IOS Press, Amsterdam, 1998, 337-342.

[23] Valentino, M.; Ruosi, A.; Pepe, G.P.; Peluso, G. *Journal de Physique IV* 1998, 249-252.

[24] Vogelesang, B.; Vlot, A. *J Materials Processing Technology* 2000, *103*, 1-5.
[25] Bonavolontà, C.; Pepe, G.P.; Peluso, G.; Valentino, M.; Caprino, G.; Lopresto, V. *IEEE Trans on Applied Superconduc.* 2005, *15*, 133-136.
[26] Bonavolontà, C.; Valentino, M.; Pepe, G.P *Supercond Sci Technol* 2007, *20*, 51-56.
[27] Schulte, K.; Baron, C. *Compos Sci Technol* 1989, *36*, 63-76.
[28] Wang, X.; Chung, D.D.L. *Composite Part B* 1998, *29B*, 63-73.
[29] Bar-Cohen, Y. *E-Journal of Nondestructive Testing and Ultrasonics* 1999, 6.
[30] Bray, D.E.;McBride, D. *Nondestructive Testing Techniques*, John Wiley and Sons, New York, NY, 1992.
[31] Pohl, J. in *Proc 7th EC NDT* 1998, 26-29.
[32] Ma, Y.P.; Wikswo, Jr J.P. *SPIE* 1993, *2001*.
[33] Placko, D.; Dufour, I. *IEEE Industrial Applic Society Annual Meeting* 1992, part II, 1676-1682.
[34] Bonavolontà, C.; Peluso, G.; Pepe, G.P.; Valentino, M. *EPJ B* 2004, *42*, 491-496.
[35] Ruosi, A.; Valentino, M.; Peluso, G.; Pepe, G.P *IEEE Trans Appl Superconduc.* 2001, *11*, 1172-1175.
[36] Ruosi, A.; Valentino, M.; Lopresto, V.; Caprino, G. *Composite Structures* 2002, *56*, 141-149.
[37] Bonavolontà, C.; Valentino, M.; Peluso, G. in *Rev Prog QNDE,* American Institute of Physics, Melville, NY, 2006, *820*, 1007-1014.
[38] Bonavolontà, C.; Valentino, M.; Pepe, G.P. *Supercond. Sci. Technol.* 2007, *20* 51-56.
[39] Bonavolontà, C.; Valentino, M.; Peluso, G.; Barone, A. *IEEE Trans. Appl. Superconductivity* 2007, *17*, 772-775.

In: Recent Advances in Non-Destructive Inspection
Editor: Carosena Meola, pp. 191-225
ISBN 978-1-61668-550-8

Chapter 7

ELECTRICAL GEOPHYSICAL METHODS TO NON-DESTRUCTIVE EVALUATION OF ARCHITECTURAL STRUCTURES

Rosa Di Maio
Department of Earth Science, University of Naples Federico II
Largo San Marcellino, 10 – I-80138 Napoli, Italy

ABSTRACT

Many efforts have been done in the last decades in non-destructive analysis of manufactures for both civil engineering and cultural heritage field. New methodological and technical approaches indeed have been developed to study architectural materials and structures. The present chapter, in particular, focuses on the electrical geophysical prospecting and its resolution power in the maintenance evaluation of artefacts as well as in the definition of the causes of eventual degradation. A description of physical, methodological and technical principles of some electrical geophysical methods is presented. Experimental tests from laboratory measurements on specimens simulating masonry structures are illustrated and discussed in order to supply guidelines for identification of the anomalies, which are due to cracks, detachments, dampness, and whose knowledge is essential for the interpretation of in situ (walls, columns, ...) micro-geophysical measurements.

INTRODUCTION

Nowadays, there exists an increasing demand of safety as well as the request of prevention and mitigation of eventual risks related to breakdown of buildings, often of high cultural and historical value. Obviously, the respect of construction standards is fundamental to prevent the risk of failures, but it does not prevent from deterioration of materials. Periodical inspections, in fact, should be made to evaluate existing conditions and discover building deficiencies at an incipient stage and to plan restoration intervents before catastrophic failure occurred.

So that, it is important to use effective non-destructive techniques able to give valuable information about the main causes of deterioration of materials and structures used in the architectural field. Generally, the degradation of architectural structures includes: variations in concrete compaction and voiding, spalling or micro-cracking in masonry, reinforcement deterioration, mat size, concrete cover, cladding attachments, metal clamps and ties, wall ties, moisture content, leaks in roofing systems, underground water infiltrations.

The attention of this capther is focused on the geophysical approach to micro-scale material evaluation that has revealed able to fulfill most of the architectural requirements of non-destructive control.

Geophysical prospecting, which basically consists in the detection of surface effects induced by natural or artificial underground sources, is largely used in all exploration fields where detailed physical and geometrical reconstruction of hidden ambients is required. The geophysical methods, indeed, are able to measure many physical quantities (potentials, currents, physical fields) by means of suitable measurement techniques, which are chosen according to the specific research purposes. In any case, their success is essentially due to the large variety of physical properties of rocks and materials, which helps to outline subsurface heterogeneities. In particular, the achievement of the geophysics in the non-destructive control of architectonic materials is strictly linked to the contrast between the physical parameters characterising sound and damaged (or repaired) parts of a structure as well as slag inclusions. The main difficulty of the geophysical prospecting in this application field concerns, essentially, the very small dimensions of the bodies to investigate, which involve low values of the signals with respect to the local noise (material inhomogeneities and environmental disturbances), and miniaturization of the probes (transmitters and receivers). To overcome these problems novel improved techniques of data acquisition, processing and interpretation have been developed.

The aim of this chapter is to illustrate the physical and technical principles of some electric-type geophysical methodologies, which have demonstrated the highest efficacy in investigating inhomogeneous micro-media, and to outline, by means of laboratory tests, how the micro-geophysics exploration can help to individuate the endogenous decay causes of architectonic materials and structures.

1. Geophysical Prospecting

Geophysics (or *Earth physics*) studies the physical properties of the Earth planet through physical measurements and methods. It is a relatively young science; in fact, although the main studies on the global properties of the Earth were performed between the XVII and XIX century, the inner structure of the Earth was recognised only in the first half of the twentieth century, using the first seismographic records. In this period, indeed, a new geophysics branch, the so-called *applied geophysics*, was born with the aim to study the physical properties and the dynamics of the shallower portion of the Earth's crust.

The term *applied geophysics* includes a wide class of non-destructive methods able to provide, through surface measurements, detailed physical and structural images of buried geological and/or anthropical structures. Such techniques are mainly devoted to [1, 2]: crustal studies (useful for hydrocarbon exploration), characterisation of shallow soils for engineering and environmental purposes, research for groundwater, minerals and other economic resources, study of space- and time-variations of physical parameters related to dynamic processes in seismic and/or volcanic high-risk areas. In the last twenty years, they are successfully applied in the cultural heritage field for reconstruction of buried archaeological remnants (see e.g. review of Linford [3] and references therein), and recently employed, more specifically, in the micro-scale inspection of architectonic structures for definition of their conservation state and the endogenetic sources of degradation (see e.g. [4÷6]).

Micro-geophysics, therefore, is a very new branch of the applied geophysics that founds on the same physical principles governing the geophysical prospecting at small-large scale, i.e. the laws of classical physics. The innovation concerns, essentially, instrumental and methodological aspects: development of high-technology equipments and of sophisticated techniques of acquisition, analysis and interpretation of data. In this context, many efforts have been undertaken to adapt and to integrate the by-now tested geophysical techniques in the micro-scale exploration (see [7÷11]).

Commonly, the geophysical methods are regrouped in two classes: active and passive methods. The first class utilises artificial sources: an electric, electromagnetic or elastic signal, generated by a transmitter device, penetrates in the subsoil and modifies on the basis of the physical and geometrical characteristics of the crossed medium. As a result, the return signals, registered at a receiver device, will contain information on the investigated buried volumes. The second group, instead, employs natural sources: electric, magnetic or gravitational perturbation fields, produced by buried features, are measured at a receiver device. Accordingly, the presence and the characteristics of the observed fields are an index of the presence and nature of the hidden sources responsible of the detected perturbations. In both cases, i.e. passive or active methods, it is possible to make another subdivision: static or dynamic methods. In the first case, distortions of a static physical field are registered in order to outline the medium responsible of such alterations, in the second case, instead, effects produced by natural or artificial sources, varying over time, are studied.

Based on the physical phenomena observed on the surface, is then possible characterise the buried structures, of geological or anthropical nature, by means of suitable physical parameters strictly linked to the different exploration methodologies, i.e. it is possible to obtain a distribution of the physical characteristics of the investigated hidden volumes. Indeed, systematic variations of these properties can define any physical or structural inhomogeneity (*anomaly*), whose knowledge is essential for both the detection of buried structures and the determination of any dynamical process to which the structures are or could be subjected over time.

Obviously, the choice of the most appropriate methodology bases on the research purposes. Due to complex phenomena linked to material's inherent heterogeneity and to interpretative ambiguities, which are present in each method, a geophysical multi-methodological approach was proved to be the best procedure to attain meaningful answers for a general characterisation of buried structures.

In particular, this chapter focuses on geophysical methodologies of electric type, since other methods based on elastic or thermal signals, like the ultrasonic or the infrared termography techniques, are largely discussed in other chapters of this volume.

The successful application of electric methods in non-destructive evaluation of buildings or historical edifices is essentially due to the large variations of the electrical parameters that characterise stones and similar architectural block devices in presence of water infiltration, micro-fracture systems, voids. Starting from the physical and methodological basis of the electric geophysical methods, some details will be provided on the operating techniques as well as on the complete explorative process, from data acquisition to data processing, representation and interpretation for micro-structural studies. The attention, in particular, will be devoted to three methodologies: direct current geoelectrics, self-potential and ground penetrating radar.

2. A Brief History of Electrical Geophysical Exploration

The electrical geophysical exploration is aimed to create an image of the underground in terms of its electrical properties. Commonly, a difference between measurements made with direct current (DC) or with alternating current is made. The methods of the first type, in fact, are generally defined as electric methods, while those of the second type, electromagnetic methods [12]. Anyway, the physical laws governing both types of methods are the well known Maxwell's equations; the so-called DC methods are merely a low-frequency asymptotic application of the fundamental equations.

The development of a pragmatic use of Maxwell's equations to explore the interior of the Earth is due to Conrad Schlumberger, Professor of Physics at the Ecole des Mines in Paris. In the period 1911-1914 he developed practical methods for measuring earth electrical resistivity in both laboratory and field experiences [13]. Following these first applications of the electrical methods at the subsurface exploration, in 1927 Saba Stefanescu, Professor at the Romanian Geological Institute, developed the mathematical basis for solution of the forward problem of computing electromagnetic field behaviour in suitable earth models [14]. The last fundamental step in the development of the theory governing the electromagnetic prospecting is due to Andrei Nikolaevich Tikhonov, Professor at the Department of Physics and Geophysics at the Moscow State University. In fact, in 1943 he published his first paper [15] on the inverse problem – determination of the earth model from a set of observations – which up until then had been considered by mathematicians to be unsolvable. Basing on practical results from a DC survey for oil in the Urals forelands, he showed that such ill-posed problems could be solved and laid the basis for a new topic in applied mathematics, the theory of solution of ill-posed set problems. Successively, many researchers have provided significant contributions to the development of electrical methods for geophysical prospecting; obviously the major revolutions have occurred in technology, particularly with the advance of digital computers and solid state electronics, which have permitted their application in many exploration fields.

A detailed chronicle of the electric methods in geophysical exploration is given in Zhdanov and Keller [12]. This reference, in particular, provides an exhaustive chronological listing of texts and monographs on the geoelectrical methods to which the reader is referred.

As previously mentioned, this chapter focuses on three electrical methodologies, of which two based on DC signals, the DC geoelectric and the self-potential methods, and the third based on alternating current, the ground penetrating radar method.

3. Direct Current Geoelectric Method

Direct current geoelectrics (DCG) is an active geophysical method aimed to define the electric resistivity values that characterise any inhomogeneous structure of physically different materials in electrical contact. Electrical resistivity of a material (or rock) states its capacity to permit the passage of electric charges and depends on many factors, such as rock- or material-porosity, presence of metal-like mineral particles, water content in fissures and fractures, saturation and ionisation degree of the pore electrolytic solution as well as its temperature. The water content is doubtless the major responsible of the electric conduction in non-metallic bodies. In dry conditions, in fact, many materials or rocks are non-conductive, i.e. they are characterise by high resistivity values, which decrease rapidly in presence of fluids at high ionic concentration. Therefore, a DCG survey is able to give not only information on the structural pattern of the underground volumes but also on their physical conditions.

Resistivity measurements are primary used for structural studies, i.e. determination of depth and thickness of geologic strata. Recent applications include also environmental, engineering and archaeological studies, e.g.:

- physical and geometrical definition of the geological structures for natural risk evaluation (seismic, volcanic and hydrogeologic);
- detection and mapping of contaminant plumes, spills and leaks;
- detection and mapping of landfills, trenches, buried waste, drums, underground structures and utilities;
- evaluation of soil and rock properties and man-made structures.

3.1. Theoretical Background

A DCG measurement is performed by inducing an electric current into the ground and recording the electrical potential arising from current input. In the common situation, the current is sent into the ground by means two electrodes (e.g. metal spikes) of opposite polarity, connected to a stationary current generator (battery), as well as the distribution of the arising potentials is obtained by measuring the difference of potential (d.d.p.) between two electrodes driven into the ground and connected to a sensitive voltmeter. Thus, the DCG method can be employed only if a satisfactory electrical contact with the ground can be realised. Figure 1 shows the pattern of the current flow and of the potential distribution for two point current sources (*A* and *B*) lying on the surface of a homogeneous ground. The dashed lines represent the paths of the electric current in going from the electrode *A* (*current source*) to the electrode *B* (*sink*), while the continuous lines represent the equipotential surfaces, i.e. curves characterised by the same value of the electric potential, indeed a voltmeter connected between any two points on such a curve would show no voltage difference. As it can be seen, every equipotential cuts a current path at right angles.

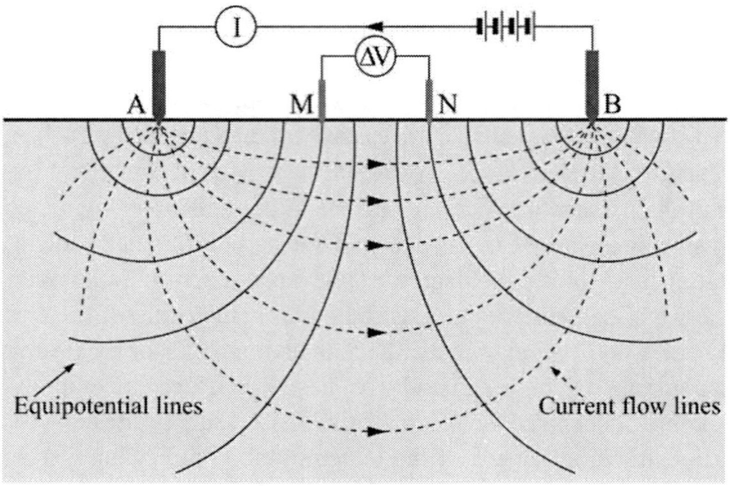

Figure 1. Current flow lines and equipotential surfaces arising from a current source, A, and a sink, B, located on the surface of a homogeneous medium. I is the intensity of the current induced by a stationary current generator (battery), ΔV is the difference of electric potential measured between the electrodes M and N.

The simplest approach to the theoretical study of DC resistivity measurements is to consider first the case of a completely homogeneous medium. The transport of electrical charges is hereby governed by Ohm's law, i.e. a linear relationship relates the applied current, *I*, to the arising electric field, ***E***:

$$\boldsymbol{E} = \rho \boldsymbol{J} \qquad (1)$$

where ***J*** (= $I/2\pi r^2$) is the electric current density on a generic half-sphere of radius *r* (see Figure 1) and ρ represents the electrical resistivity of the homogeneous medium (the ρ parameter is a material constant and is independent by the geometry of the body).

Due to the stationary conditions, from Maxwell's equations follows that the electrostatic field (i.e. generated by an immobile charge distribution) can be expressed in terms of the gradient of a scalar potential function, viz.:

$$E = -\frac{\delta V}{\delta r} \qquad (2)$$

where *V* is the electrostatic potential. Therefore, using Eq. (1), the electrostatic field can be expressed as:

$$E = -\frac{\delta V}{\delta r} = \rho J = \rho \frac{I}{2\pi r^2} \qquad (3)$$

From Eq. (3) is simply now to derive the electric potential, *V*, induced by the applied current, *I*, in a generic point of the surface at a distance *r* from the current point source:

$$V = -\int E dr = \frac{I}{2\pi}\rho \int \frac{1}{r^2} dr = \frac{I\rho}{2\pi r}. \tag{4}$$

An exhaustive description of the theory for the static electromagnetic geophysical methods is provided, e.g., in [1, 12].

Let us consider now a generic configuration of the four electrodes on the ground surface (Figure 2); if r_1 and r_2 are, respectively, the distances of the potential electrode *M* from the current source *A* (+*I*) and the sink *B* (-*I*), due to the additive property of the potentials, from eq. (4) follows that the potential in *M* due to the current flow from *A* to *B* is given by:

$$V_M^{AB} = \frac{I\rho}{2\pi}\left(\frac{1}{r_1} - \frac{1}{r_2}\right). \tag{5}$$

Moreover, if r_3 and r_4 are, respectively, the distances of the potential electrode *N* from the positive (*A*) and the negative source (*B*), from Eq. (5) follows that the d.d.p. between the surface points *M* and N, creates by the current induced in the medium through the electrodes *A* and *B*, is given by:

$$\Delta V_{MN}^{AB} = \frac{I\rho}{2\pi}\left[\left(\frac{1}{r_1} - \frac{1}{r_2}\right) - \left(\frac{1}{r_3} - \frac{1}{r_4}\right)\right] = \frac{I\rho}{2\pi}\left(\frac{1}{r_1} - \frac{1}{r_2} - \frac{1}{r_3} + \frac{1}{r_4}\right). \tag{6}$$

Eq. (6) permits to retrieve the expression of the electrical resistivity, that is:

$$\rho = \frac{\Delta V}{I} \frac{2\pi}{\left(\frac{1}{r_1} - \frac{1}{r_2} - \frac{1}{r_3} + \frac{1}{r_4}\right)}. \tag{7}$$

Putting now:

$$k = \frac{2\pi}{\left(\frac{1}{r_1} - \frac{1}{r_2} - \frac{1}{r_3} + \frac{1}{r_4}\right)} \tag{8}$$

we obtain:

$$\rho = k\frac{\Delta V}{I}, \tag{9}$$

which represents the resistivity of the considered homogeneous half-space. K is the so-called geometrical factor, since it is a function of the chosen electrode configuration. The measure unit of ρ, in the MKS system, is Ohm•m (Ωm).

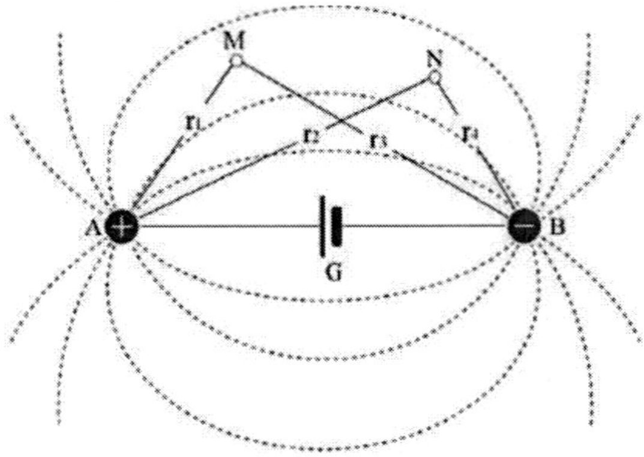

Figure 2. Current distribution (dashed lines) in an electrically homogeneous medium (plan view).

In presence of inhomogeneous medium (e.g. layered media, cavities, …), that is the case of real situations, the distribution of the electric potential will no longer be spherical (Figure 3), so that Eq. (1) is no longer valid.

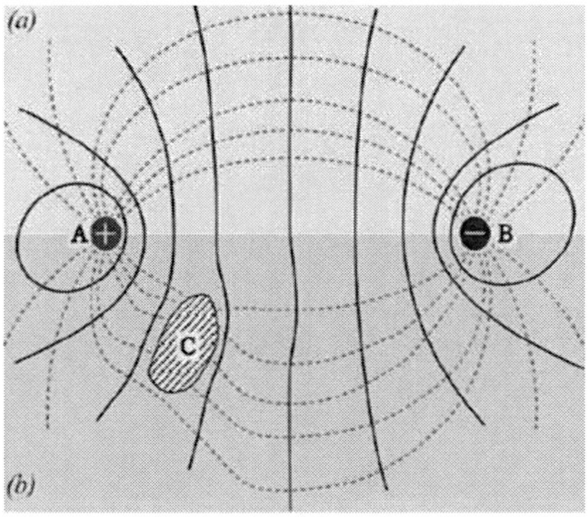

Figure 3. Plan view of the current (dashed lines) and equipotential (continuous lines) distribution in a homogeneous (a) and inhomogeneous (b) medium. C simulates a resistivity inhomogeneity (conductor body).

In this case, the resistivity value given by the Eq. (9) represents an *apparent* resistivity value, ρ_a, integrated over the different layers (or bodies) that are influenced by the induced current, therefore, it has no longer a direct physical meaning. The apparent resistivity, in fact, is defined as the real resistivity of a fictitious, homogeneous and isotropic medium, in which, for a given electrode configuration and intensity current I, the measured electric potential difference, ΔV, is equal to the value of the inhomogeneous medium. Consequently, the resistivity measured on the surface of an inhomogeneous subsoil will vary at varying of the mutual disposal of the electrodes, i.e. each observed resistivity represents an apparent resistivity value. In such a case it is necessary to invert the data in order to obtain the real resistivity values characterising the buried structures.

3.2. Acquisition and Interpretation of DC Resistivity Data

The standard field procedure of the DC geoelectrical method consists in the determination of an as dense as possible set of apparent resistivity values in correspondence of as many different positions of the electrode device, which is used to produce the input current field, by means a stationary current generator, and to detect, at a high-input impedance voltmeter, the electrical response of stimulated underground materials.

Three are the most commonly used electrode configurations: Wenner, Schlumberger and Dipole-Dipole (Figure 4). The first two, which take the name from their originator, Frank Wenner [16] and Conrad Schlumberger, are also defined as symmetric arrays, indeed the four electrodes, *A* and *B*, *M* and *N* are collinear and located symmetrically with respect to the mid-point of the array (see Figure 4). For the Wenner array, the electrodes are equidistant (*AM*=*MN*=*NB*), while for the Schlumberger configuration the *AB* distance is much larger than the *MN* distance. Finally, for the dipole-dipole array, the inter-electrode distance for each couple of electrodes is insignificant with respect to the distance between the centres of the two dipoles.

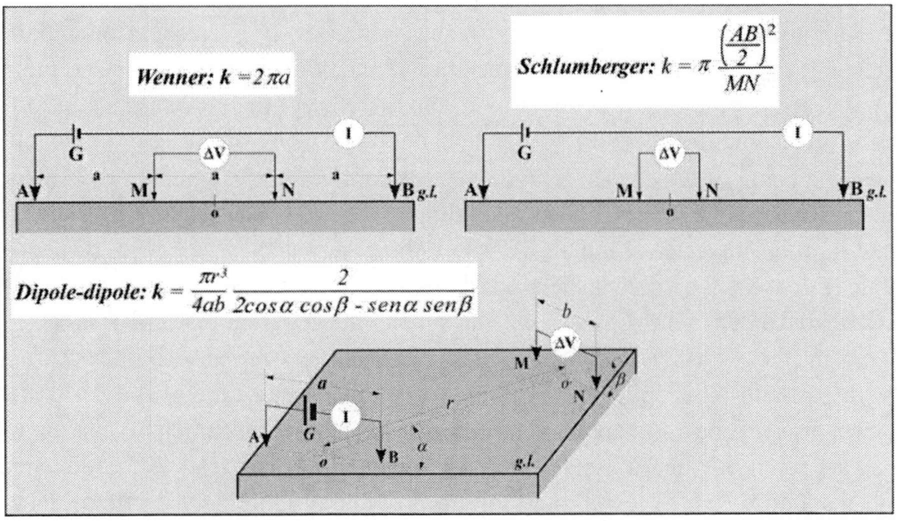

Figure 4. Electrode arrays commonly used in the electrical prospecting.

The geometric factor, *k*, for each array is indicated in Figure 4. It can be easily obtained substituting to r_1, r_2, r_3 and r_4 of Eq. (8), respectively, the distances *AM*, *MB*, *AN* and *NB* corresponding to each configuration. For the Schlumberger array, in particular, is indicated the approximate geometric factor, related to the approximation that the *MN* spacing is much smaller than the *AB* spacing (generally $MN \leq 1/5AB$). In this approximation, the d.d.p. observed between *M* and *N* roughly corresponds to the electric field in the mid-point of the *MN* spacing (see [12]).

In field procedures, the choice of the array is mainly related to the space available to lay out the array and the required maximum investigation depth [17] as well as to its sensitivity to lateral inhomogeneties [18]. A general guide to the suitability of the quoted electrode configurations is given in Reynolds [2].

3.2.1. Data Acquisition

In the past, the most commonly used technique in the DCG prospecting is doubtless the vertical electric sounding (VES). It is based on the assumption that the resistivity, ρ, varies only with the depth, z, i.e. it is assumed that the electrical structure of the earth can be described with a one-dimensional resistivity function. This assumption is equivalent to admit that the resistivity lateral changes are so slow as to not affect $\rho(z)$ at any one location of the sounding. Then, the lateral variations can be determined with a series of vertical electric soundings made along a profile or over an area and interpreted as a set of slowly changing $\rho(z)$ functions.

As the depth to which the current penetrates depends on the current electrode distance, technically a VES is realised by increasing the distance between the current electrodes and keeping fixed the centre of the array that corresponds to the location of the sounding. Specifically, in the case of the Wenner array, all four electrodes have to be moved to new positions as the inter-electrode distance are increased, while for the Schlumberger configuration, the potential electrodes are placed at a fixed spacing and the current electrodes are placed at progressively larger distances (Figure 5a). When the voltage measured between M and N falls to very low values (owing to the progressively decreasing potential gradient with increasing current electrode separation), the M and N electrodes are spaced more widely apart. Finally, if a dipole-dipole array is used, the increase of the current penetration depth is obtained by increasing the inter-dipole distance, not by growing the current dipole length.

For each spacing of the chosen array, a measure of the I and ΔV quantities is made and an apparent resistivity value is calculated according to Eq. (7). The apparent resistivity values are plotted in a log-log scale (Figure 5b), where the x- and y-axes represent, respectively, the logarithmic values of the current electrode half-separation for the symmetric arrays, or the dipole-centre distance for the dipole-dipole array, and the apparent resistivity values.

At present, the traditional vertical soundings are largely replaced by new computerized methods. In particular, the application in the last twenty years of the geophysical methods to near-surface prospecting (archaeological, engineering and environmental) and/or to micro-scale analysis (civil engineering and cultural heritage field) has lead to the development of ultra-high resolution (u.h.r.) techniques able to provide an extremely detailed picture of the electric resistivity behaviour of the underground.

Among these methods, the geoelectrical pseudosection technique along a selected profile (also called bi-dimensional electrical resistivity tomography, *2D ERT*) is surely the most indicate for the detection of targets of very small dimensions and/or of anomalous features ascribable to causes of degradation (e.g. humidity, micro-fractures, voids). In fact, following Worthington [19], the u.h.r. geoelectrical pseudosection can be classified as a first-order 2D tomography, since it can provide in a truly objective way, and prior to any data inversion process, symptomatic resistivity anomaly images across the vertical plane through the profile.

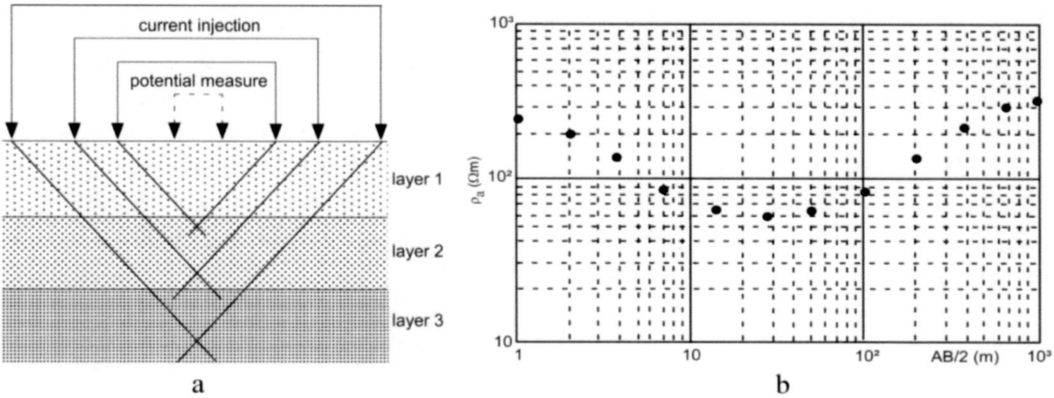

Figure 5.(a) Schematic representation of the VES technique over a three-layer stratified subsoil with different electrical properties; (b) example of VES diagram showing apparent resistivity, ρ_a, vs. current electrode half-spacing for a three-layer medium with $\rho_1 > \rho_2 < \rho_3$.

The development of the ERT method has taken place since the 1980'ies [20, 21]. The data collection is generally done with multi-electrode systems: a large number of electrodes are placed on a line at equal distance; the electrodes are connected to a switching unit and to a resistivity-meter by multicore cables. Figure 6 schematically depicts the field procedure of the ERT technique along a profile using a Wenner array with basic distance *a*. The electric current is sent into the ground by means the two current electrodes *A* and *B*, and the generated voltages are measured at the two potential electrodes *M* and *N*. A computer, typically build up into modern resistivity-meters, controls which electrodes are used as current and potential electrodes. In surveying, several traverses are performed with various values of *n* (1, 2, ...), that is an integer number of times the spacing *a*. The apparent resistivity values, calculated using the Eq. (9), are assigned, for each horizontal axis *n*, at the intersections of two converging lines at *45°* from the considered couple of current electrodes. The distribution of the apparent resistivity values along the section is then submitted to a comparative analysis in order to determine sectors (or layers) characterised by the same apparent resistivity value.

The resolution power of the pseudosection depends, obviously, on the density of the data points along the section, correlated to both the number of channels of the resistivity instrument and the basic electrode spacing. In fact, the smaller the electrode spacing the smaller the distances that the field has to walk, permitting to outline possible very small structural and/or physical discontinuities. In micro-scale surveys, an initial electrode spacing of fewer centimetres is very common, since of the order of the horizontal dimension of the expected anomalies. So as, a maximum electrode spacing of about *50 cm*, with a total profile length of *150 cm*, may permit an investigation depth of about *20-30 cm*, that is of the order of the masonry thickness mainly exposed to degradation risk or reinforcements. If the suspected anomalies are of greater dimensions or a major investigation depth is required, the above quoted parameters will change accordingly.

Since all geological or anthropical structures are 3D in nature, a 3D mapping should in theory give the most accurate results. A 3D image of the study area can be obtained by *i)* performing a large number of 2D tomographies along parallel profiles and processing the data according 3D modelling, *ii)* carrying out a 3D survey. This last is achieved through a multicore cable that connects a series of electrodes, usually arranged in a square (or rettangular) grid, that are used for both the input current and the measure of the potential difference [22]. The method consists in lighting the underground through the current electrodes, as in the 2D tomography, but the potential measurements are carried out along the two directions (or the diagonals) of the grid. The measurement sequence is defined by the operator using suitable programs that permit the optimization of the data acquisition sequence in order to reduce the survey time. For a given area, in fact, the 3D survey cost is comparatively higher than a 2D survey due to the time required for both the acquisition of very large data sets and the data processing. Figure 7 shows an example of 3D electrode arrangement for an electrical tomography on a column [5].

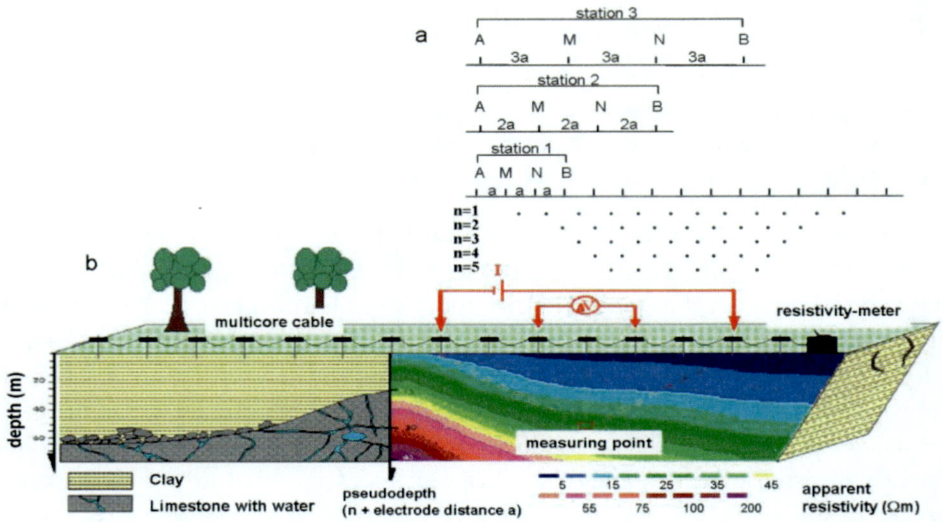

Figure 6. (a) Scheme of ERT data acquisition using a Wenner array with electrode distance a; (b) a synthetic subsoil model (left) and the corresponding resistivity pseudosection (right).

Figure 7. Example of electrode arrangement for a 3D ERT survey on a column (after [5]). The current is injected through two electrodes: one at the top and the other at the base of the column.

3.2.2.Data Interpretation

As shown in the previous section, no DC geoelectrical measurement is able to give, whatever the chosen technique (vertical sounding or tomography), a direct information about the real resistivity distribution in the subsurface (or inside any masonry structure) unless it is homogeneous. To build the real resistivity distribution of an inhomogeneous underground, it is necessary to numerically transform or invert the apparent resistivity values into real resistivities.

Interpretation of experimental VES curves is performed generally by means a 1D direct (or *forward*) interpretation method. The field curves of apparent resistivity, indeed, are compared with a set of apparent resistivity theoretical curves coming from the solution of the forward problem for an one-dimensional horizontally stratified earth model [12, 23, 24]. The match gives the thickness and the electrical resistivty of the hypotised horizontal layers. In 1971, Ghosh [25] proposed a convolution method to compute resistive curves for layered earth structures by which computers can be used [26].

The true resistivity distribution coming from 2D or 3D resistivity tomographies can be derived using a forward approach or as solution of the so-called *inverse* problem, which is in principle unique in absence of disturbances of any kind. The algorithms most commonly used for the inversion of 2D or 3D resistivity tomography data refer to a smoothness-constrained least squares method [e.g. 27÷32] based on a linearisation process of the apparent resistivity values as function of the involved real resistivities. Basically, the inversion procedure founds on a finite element forward modelling of the investigated buried volumes by means a discrete number of small blocks each characterised by a constant resistivity value (solution of the *forward* problem). Then, the application of optimization iterative algorithms (*inverse* problem) attempts to determine the resistivity values of the blocks so as to minimize the difference between the calculated and the observed apparent resistivity values. In 3D-inversion scheme, a Lagrangian multiplier is also used in order to smooth the final inverted model. The fit goodness is expressed in terms of the root mean square (*RMS*) error.

4. Self-Potential Method

Self-Potential (SP) method is a passive geoelectrical method used to outline underground anomalous electric charge polarizations, which can be measured between any two points on ground's surface. Spontaneous potentials were experienced for the first time in 1830 by Robert Fox [33] in a sulphide deposit area (Cornwall, England). Indeed, he observed that even when no current was flowing through the subsoil, the voltmeter detected a potential difference when the measuring electrodes were connected to a galvanometer.

Since 1920, the SP method has been used in mining exploration, while over the last thirty years has been successfully applied in many others application fields, such as hydrogeological and geothermal exploration, engineering and environmental studies, natural risk assessment (volcanologic, seismologic, landsliding). Recently, its use is involving also the archaeological and architectural prospecting.

4.1. Origin of Self-Potentials

Self-potentials can be generated by many natural sources thought the main factor responsible for all is groundwater. The underground fluid circulation, in fact, may originate many electrochemical phenomena, which give rise to the primary sources of the natural potentials (or, equivalently, of the natural electric field) observed on the ground surface. In addition, spontaneous polarisation can be also generated by the geometry of buried geological or anthropical structures, since secondary sources can be also induced across the resistivity discontinuity surfaces by primary electric charges.

Generally, the main electrochemical processes responsible of the SP anomalies are subdivided in two groups: mineralization potentials and background potentials. These last commonly divided in *i)* electrochemical potentials and *ii)* electrofiltration potentials [1, 34]. Of course, in mining exploration the mineral potentials represent the main SP anomaly sources and the background potentials constitute noise sources; contrarily in geothermal, groundwater, natural risks, archaeological or architectural prospecting, the background potentials are the principal sources to investigate.

Mineralization potentials (exclusively negative) occur in the vicinity of sulphides, oxides and graphite deposits, i.e. when an underground metallic body is present (Figure 8a). Although a model able to explain all the occurrences of mineral potentials does not exist, to date the most exhaustive explanation of this electrochemical phenomenon is that proposed by Sato and Mooney [35]: when a metallic body is partially immerged in an ionic conductor (water table), a cathodic electrochemical half-cell forms by the chemical reduction of the ions in the surrounding electrolyte, i.e. they gain electrons. Contrarily, below the water table, an anodic electrochemical cell operates in which oxidation phenomena are dominant and ions lose electrons. The massive orebody has the role to permit the flow of electrons from the lower-half of the body to the upper-half. Hence, the upper surface becomes negatively charged (a negative anomaly is observed on surface) and the lower half becomes positively charged.

Electrochemical potentials (Figure 8b) are very likely due to differences in the mobilities of electrolytes having different ionic concentrations within ground water. An electrochemical potential can also be observed when two electrodes are immersed in a homogeneous solution, but the ionic concentrations are locally different (Nerst or shale potential). This is a special case of the processes describing the diffusion potentials. Since, as it is well known, the ionic mobility is also directly dependent on the temperature, the higher the temperature and the concentration differences, the larger the electrochemical potentials will be. Hence, the electrochemical potentials represent the main SP anomalies observed in geothermal and volcanological areas.

Electrofiltration (or *streaming*) *potentials* (Figure 8c) are observed between the terminals of a capillary or a porous medium permeated by an electrolytic fluid (see [36]). Such potentials are due to selective adsorption of ions on the walls of the capillary, sometimes combined with a not uniform velocity distribution of positive and negative ions. The intensity of the observed potential drops is directly proportional to the pressure gradient (responsible of the fluid circulation), the electrical properties of the electrolyte as well as the rock pore network through which the electrolyte flows, and inversely proportional to the fluid viscosity. The formation mechanism of these potentials can be ascribed to the presence of dissolved salts in the underground fluids, which make them rich in anions and cations. The free liquid in the centre of the rock pore is usually enriched in ions with a specified sign, while the ions of opposite sign are usually adsorbed on the solid surface. The free pore water carries an excess of charge of one sign, part of which accumulates at the solid-liquid interface forming a stable double-layer. When the liquids are forced through the porous medium, owing to the action of the pressure gradients, the water molecules carry along with them free ions in the diffusion part of the pore. This relative movement of cations with respect to the firmly attached anions generated the well known *streaming potential* [37].

Electrokinetic phenomena are the main responsible of potential anomalies in volcano-geothermal areas [38÷43], seismically active structures [44÷49] and archaeological sites [7, 50÷53]. At present, the study of these potentials seems to provide new insights of the SP method at a micro-scale analysis of architectural materials and structures. Indeed, loose stones or similar architectural block devices, characterised by a dense micro-fracture network permeated by fluid circulation (moisture), may give rise to SP anomalies [4, 9, 54] indicative of the degradation state of the investigated structures.

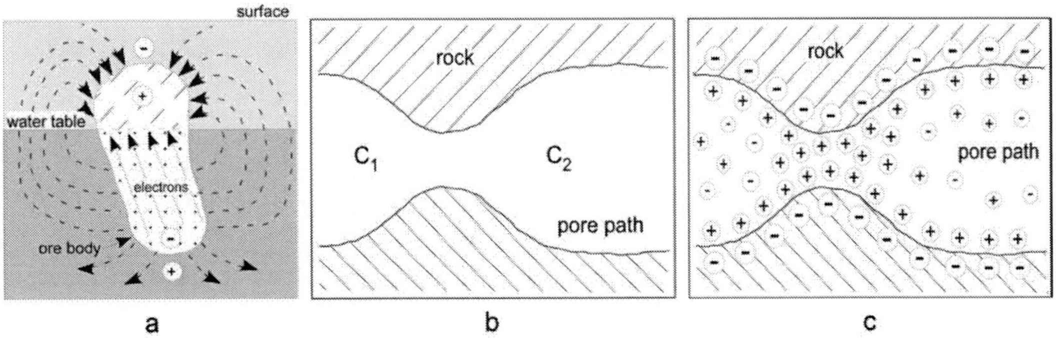

Figure 8. Schematic representation of mineralisation (a), electrochemical (b) and electrofiltration (c) self-potential source.

4.2. Acquisition and Interpretation of Self-Potential Data

The standard field procedure of the SP method consists in measuring variations in natural potentials along profiles or areas. The aim is to obtain a SP profile (Figure 9a) or a SP contour map (Figure 9b) outlining the anomalous zones which are connected with underground anomalous electric charge concentrations.

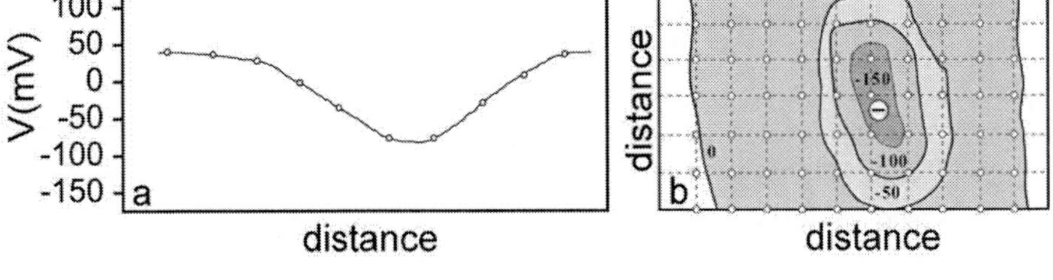

Figure 9. Self-potential anomaly due to the orebody shown schematically in Figure 8a observed along a profile (a) or over an area (b) (contour lines in mV).

Among the surface geophysical techniques, the SP method is considered as the cheapest in terms of equipment and the simplest in term of field procedure. Nevertheless, the SP data collection requires some carefulness to be taken into account during the surveying.

The SP equipment includes two electrodes, a high input impedence multimeter (≈ 10 GΩ), electrodes-multimeter connecting wires, a PC to collect and to average the data set sampled at a fixed frequency over a pre-established period. The frequency sampling as well as the acquisition interval length are established according to the test-site noisy. For most the SP applications, non-polarisable electrodes are commonly used. They are made of a copper rod dipped in a saturated solution of copper sulphate which can percolate through the porous base in order to make electrical contact with the ground. Zinc- or silver electrodes in saturated, respectively, zinc sulphate or silver chloride solution can be also used [2]. In non-destructive control of manufactures, instead, electrodes commonly used in the medical field are utilised (see Section 6.1.1). They, in fact, allow to overcome some of the problems (perforation, large dimension, high-contact resistance, signal instability, etc.) of the conventional electrodes.

4.2.1. Data Acquisition

Two measurement techniques are commonly used in the SP prospecting: gradient and total field. The gradient technique employs two electrodes moving along a selected profile (2D survey) or a grid (3D survey); the SP measure is attributed to the mid-point between the two electrodes and is divided by the electrode spacing to give a potential gradient (mV/m). Also the total field technique utilises two measurement electrodes, but one electrode is keeped fixed at a base station, while the other electrode is moved along the survey profile or grid. The potential gradient technique is faster and more practical than the total field technique with regard to data acquisition. Moreover, it is more sensitive to changes along the surface and lateral inhomogeneities than the total measurement. Therefore, the gradient technique is generally preferred for small- or micro-scale investigations (like archaeological or architectural prospecting), as very shallow physical or structural inhomogeneities are usually observed in soils hosting anthropical structures as well as significative changes in the shallower portion of architectural manufactures (wall, colomns,..) can be outlined due to the presence of moisture, micro-cracks and/or restored areas.

However, the gradient technique has some disadvantages as concerns the acquired data quality that may be influenced by different noise sources: electrode polarization, drift effects, time-varying potentials, soil contact effects and reading errors [55]. Other noise sources that may significantly affect SP data, whatever the adopted technique, are topographic effects, daily temperature changes, etc.. They are removed from the acquired data by suitable correction processes (see e.g. [2, 43, 51]).

As previously said, the SP measurements may involve bi- or tri-dimensional surveys. In 2D surveys, the SP measurements are collected (with gradient or total field technique) along a profile and are generally devoted to outline lateral inhomogeneities. This implies that the measure profile should be perpendicular to the suspect anomalous body, otherwise no lateral polarisation contrasts are observed. Generally, more and more profiles, diversely oriented in the survey area, are investigated. The electrode spacing as well as the profile length are chosen according to depth and size of the target. Generally, the maximum exploration depth is about 10 times the electrode separation, depending on the nature of the overburden [2], while the profile length should be at least about 5-6 times the target size. The 2D technique is generally used for large surveys areas, i.e. when a 3D exploration is difficult for both logistical problems (irregular topography, high urbanization) and large length of the connecting wires to the measuring unit if the total field technique is used.

In small- and micro-scale prospecting, the SP data are generally collected using a 3D survey procedure, which is both fast and of high-resolution. In 3D surveys, the SP measurements are performed along a grid of squared or rectangular cells of very small dimensions (of the order of few squared meters for archaeological or engineering investigations and of few squared centimetres for manufacture analysis). In these application fields, the SP anomalies are generally low in amplitude (few tenths of mV at least).

4.2.2.Data Interpretation

Field data coming from 2D or 3D SP surveys, corrected for drift and/or spurious effects, are firstly interpreted qualitatively; in fact, the SP profile shape, the contour pattern of the SP isolines, the amplitude and the polarity of the anomalies may give useful indication on the dimension and deepness of the anomaly source. For example, a long wavelength anomaly could indicate the presence of a relatively deep body as well as a high-frequency spatial anomaly of low amplitude could represent the response of a relatively shallow body interested by not significant polarisation processes, that is indicative of underground poor fluid circulation. Moreover, the presence of a couple of anomalies of opposite sign very likely indicates the contact between two different media, so-called axis of polarization, as well as the polarisation angle, i.e. the angle between the horizontal surface and the polarization axis, determines the polarisation direction of the body.

The second step of the interpretation process is of quantitative type and consists on the definition of the SP source parameters, like deepness, dimension and nature. The quantitative interpretation can be realised with both direct and inverse procedures. The direct or forward approach, as for the DC resistivity method (see Section 3.2.2), consists on calculate the SP theoretical anomalies generated by bodies of known geometry, such as point current, horizontal line, spherical, cylindrical and vertical dipole sheets, with an assumed direction of polarisation [56÷59]. Drahor et al. [51] for example have interpreted burial archaeological structures, generally found in vertical and horizontal forms in the soil, by assuming an inclined sheet model. In the last decades, several inversion techniques have been developed (see e.g. [60÷62]). The proposed approaches fundamentally base on an initial estimate for the model parameters and then modified by an iterative process. In each iteration, a comparison between the theoretical anomaly and the experimental data is made by calculating the RMS error. The inversion procedure is continued until the RMS error is less or equal than an allowable error. Jagannadha Rao et al. [62], in particular, utilise Marquardt's optimization procedure for parameter's estimation [63].

Interpretation of SP data coming from micro-scale surveys on architectural materials and structures is generally not easy, due to the nature and the complex distribution of the sources (micro-cracks, voids, moisture) very likely responsible of the observed SP anomalies. The application of the SP method to micro-scale manufactures analysis would has perhaps never been utilised in the past, essentially for the lack of a reference phenomenological model and of a suitable 3D representation of the underground charge polarisation centres. Recently, by a conceptually simple adaptation of the water circulation/charge polarisation field phenomenology [42] to the scale of stones, columns and walls, using the newly developed data acquisition techniques and data tomographic processing methods [64, 7], have been put in evidence SP micro-anomalies, which have opened new vistas in the conservation state diagnosis of any architectural structure [4, 54]. In particular, the quoted SP tomographic inversion scheme aims to the definition of the underground *electric charge occurrence distribution* by means the decomposition of a synthetic SP surface field into a sum of elementary contributions due to a discretised distribution of charge accumulation centres. The inversion problem consists in recovering the underground discretised charge distribution, responsible of the measured SP field on surface, through a normalised cross-correlation procedure between the observed natural electric field and the synthetic SP field. This procedure properly defines the *charge occurrence distribution* (COD) *function*. Using this physical scheme, the tomographic process consists of scanning the whole lower half-space (the tomospace) along a sequence of horizontal planes parallel to the field plane, for any fixed depth. For each depth, the complete grid of regularly distributed COD values in the horizontal plane can be utilised to draw contoured slices in order to single out the zones of highest concentrations of polarised electric charges. Due to the normalisation procedure, the COD *function* is constrained to vary inside the interval [-1, +1]: positive values are the result of a major influence from positive charge accumulations, while negative values result from negative charge concentrations.

5. The Ground Penetrating Radar Method

Ground Penetrating Radar (GPR) is an active geophysical method aimed to outline the sharp changes in the electromagnetic (*em*) properties of the materials by emission of radio-frequency electromagnetic waves. The *em* characteristics of a medium mainly depend on its composition and water content, both, in fact, control the speed of radiowave propagation and the attenuation of the electromagnetic waves goes through it. In particular, the radiowave velocity depends on the light velocity in the free space, the electrical conductivity, the relative electric permittivity and the relative magnetic permeability. These two last representing the extent to which a material concentrates electric and magnetic flux, respectively. Whereas, the wave attenuation is linked to the capability of the medium to transmit the radiowaves: water-saturated materials, for example, can be considered as opaque to radiowaves owing their properties to adsorb or reflect the wave energy, conversely, materials, such as polar ice, are virtually transparent to radiowaves. The success of the GPR method founds on the contrast between the relative dielectric constants characterising different materials in electrical contact, which are responsible of the reflection of the incident electromagnetic wave: the higher the contrast, the greater the radiowave energy reflected.

The first application of electromagnetic waves for detecting buried objects is due to Hülsmeyer in a German patent in 1904, also if the first paper on such investigations goes back to Leimbach and Löwy [65], which used a system based on continuous wave transmission. Pulsed radar signals were used for the primary time in 1926 by Hülsenbeck [66] to investigate the subsoil properties. Development of pulsed techniques occurred substantially over the following five decades [2]. The primary use of the GPR method to geological application was in the 1960s for glaciological studies. An useful review of glaciological applications of radar is given in Bogorodsky et al. [67]. The suitability of the technique to a much wider range of applications, such as fracture detection in rock salt, definition of soil and rock layer thickness, assessment of shallow deposits, has followed (see e.g. [68÷72]). Over the last three decades, moreover, the GPR methodology has been largely used as a rapid, non-destructive and non-invasive technique for civil engineering, archaeological and planetary purposes, such as underground shallow cavities and utilities detection (see [73, 74]), road pavement analysis [75, 76], identification of potential archaeological excavation areas (see i.e. [7, 77÷81]), characterisation of Martian soils analogs for Mars exploration rover mission (see [82, 83]).

Finally, in the last decade a large utilisation of the GPR methodology to non-destructive analysysis of engineering and architectural materials and structures has been experienced. This includes diverse features, such as definition of layer thickness (e.g. [84, 85]); characterisation of different constructive materials (e.g. [86]); identification of voids and/or degraded zones (e.g. [87÷90]); water content mapping (e.g. [91, 92]); location of reinforcing bars and metal elements in concrete structures (e.g. [93÷95]).

5.1. Theoretical Background

A GPR measurement is performed by inducing short electromagnetic pulses (i.e. waves at radar frequencies: 10 MHz - 2 GHz) in the ground and recording the amplitude and the delay time of reflections from physical and/or structural discontinuities inside the investigated volume. These reflections, in fact, originate whenever the GPR impulse comes across any structure (such as shallow layers, cavities) or discontinuity in its physical properties (such as water content, porosity) showing a conductivity (σ) and/or a relative permittivity contrast (ε_r) with the surrounding medium (Figure 10). As previously mentioned, these properties strongly influence the underground propagation of radiowaves, i.e. their velocity and attenuation.

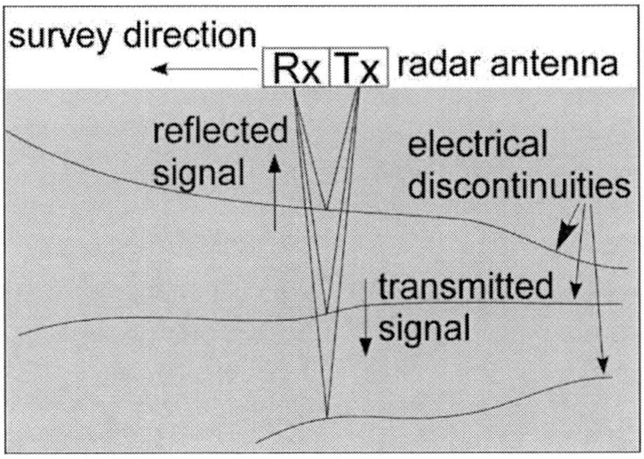

Figure 10. GPR operating principle.

Since the radar pulses are electromagnetic waves, their propagation is described by the Maxwell's equations, in particular it is proved [96] that the velocity and the attenuation factor of an electromagnetic wave of frequency f are given, respectively, by the following equations:

$$v=\frac{c}{\left\{\frac{\varepsilon r\,\mu r}{2}\left[1+\left(\frac{\sigma}{\omega\varepsilon}\right)^{2}+1\right]\right\}^{1/2}} \tag{10}$$

$$\alpha = \omega \left\{ \frac{\varepsilon\mu}{2} \left[1 + \left(\frac{\sigma}{\omega\varepsilon} \right)^2 \right]^{1/2} - 1 \right\}^{1/2} \tag{11}$$

where c is the light velocity, μ_r ($= \mu/\mu_0$) is the relative magnetic permeability, ε ($= \varepsilon_r\varepsilon_0$) is the electric permittivity and ω ($= 2\pi f$) is the angular frequency of the *em* wave. μ_0 *and* ε_0 represent, respectively, the magnetic permeability and the electric permittivity of free space.

Since the majority of the materials under consideration during a GPR survey have magnetic permeability values close to that of the free space, for most practical applications μ_r is considered equal to 1.

In presence of low-loss media, where the GPR method is applied successfully, the ratio $\sigma/\omega\varepsilon$, that compares in Eqs. (10) and (11), is about zero, so that the wave velocity is given by:

$$v = \frac{c}{\sqrt{\varepsilon_r}} \tag{12}$$

commonly used in practical field measurements. Moreover, for these media, it can be shown [96] that the reflection coefficient, R, i.e. the amount of energy reflected from a discontinuity separating two layers with different electric permittivities, may be determined according to the following equation:

$$R = \frac{\sqrt{\varepsilon_1} - \sqrt{\varepsilon_2}}{\sqrt{\varepsilon_1} + \sqrt{\varepsilon_2}} \tag{13}$$

where ε_1 and ε_2 represent, respectively, the electric permittivity of the first layer and the adjacent layer. It is worth to point out that the R parameter defines both the rate of reflected energy, collected on surface by the receiver, and the remnant energy, *(1-R)*, transmitted by the discontinuity to deeper portions of the subsoil.

The incident energy amplitude, E_0, instead, depends on the electrical properties of the materials:

$$E_x = E_0 e^{-\alpha x} \tag{14}$$

where E_x is the field strength at a distance x from the source and α is the attenuation parameter (see Eq. (11)).

From Eqs. (14) and (11) it follows that high conductivity bodies and/or high antenna frequencies will produce large attenuation. In field practice, the transmitter impulse generally includes a broad frequency interval allowing that some of the source energy penetrates to a reasonable depth within the soil (see [96, 97]).

5.2. Acquisition and Interpretation of GPR Data

5.2.1. Data Acquisition

The standard field procedure of the GPR method is based on the measurement along a profile of travel time and reflection amplitude of a short electromagnetic pulse transmitted through the surface and then partly reflected from electrical interfaces within the subsurface. Therefore, the GPR systems commonly consists of three components: a pulse generator producing a single pulse of a given centre frequency and power, transmitting (T_x) and receiving (R_x) antennae, which, respectively, transmit the pulses into the medium and capture the reflected signal, and a recorder that samples and storages the reflected signals. The operating mode is either monostatic or bistatic: in the first case one antenna is used as transmitter and receiver, in the second case two antennae, transmitting and receiving, are used. Since the velocity of the radiowaves is very high (of the order of few meters for nanoseconds, *ns*), the radiowave travel time, i.e. the time space between the wave transmission and its return to the receiving antenna, ranges from few tens to some thousand of nanoseconds. Hence, a high accuracy system is required for measuring the transmit instant and a very short transmitted pulse length (generally < 20 ns) has to be used for providing resolvable reflections (see [2]).

The vertical resolution of a GPR survey is essentially linked to the wavelength of the energy transmitted in the underground: the higher the antenna centre frequency the greater the vertical resolution, but the lower the exploration depth. Table 1 reports, indicatively, the maximum interface depth detectable with GPR antennae for common geological and man-made materials. Generally, for geological applications, antennae with centre frequencies less than 250 MHz are used, while antenna frequencies greater than 1000 MHz are utilised for engineering or architectonical analyses. The horizontal resolution of a GPR measurement instead depends on the radiation pattern of the incident energy in the underground. For most antenna, it is like a cone spreading with depth to illuminate an increasing 'footprint' of the subsurface (Figure 11a).

The collected data are generally represented in terms of a two-dimensional map, identified as radargram (Figure 11b), with the horizontal axis indicating the location of the radar antenna on the ground surface and the vertical axis representing the increasing time-delay from the initial impulse, which allows to easily visualize the underground radar anomalies. Due to the radiation pattern of the transmitted energy (see Figure 11a), a buried reflective target will be illuminated both before and after the antenna has passed over its location on the surface. This results in the common observation of point source reflectors being reproduced as hyperbolas within the data from which the subsurface velocity may be estimated from the slope of the hyperbola tails [3].

Table 1. Approximate values of the maximum penetration depth of GPR signals as function of the transmitting antenna frequency for common geological and man-made materials.

Antenna centre frequency (MHz)	Max penetration depth (m)
2000	0,40-0,60
1000	0,80-1,00
500	1,50-2,00
250	3,00-4,00

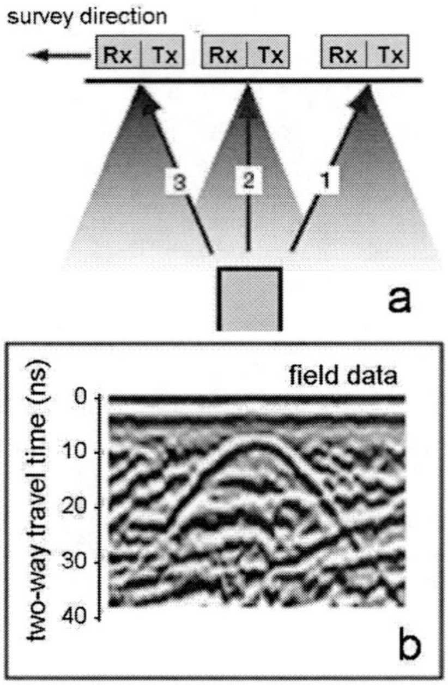

Figure 11. (a) Conical spreading of a GPR beam; (b) radargram, i.e. representation of the receiver signals as a function of their two-way travel time (figure adapted from [3]).

The successful of the GPR to engineering, archaeological and, more generally, to non-destructive testing applications has necessitated the use of 3D acquisition techniques able to provide very-high horizontal and vertical resolution. A 3D GPR survey is realised by collecting the data along parallel profiles with a spacing interval of tens or few centimetres, respectively, for engineering/archaeological or material analysis purposes [79, 97÷100]. One of the most useful representations of 3D survey data is to display them in horizontal maps of recorded reflection amplitudes (Figure 12). These maps, so-called time-slice amplitude maps, are commonly realised at various reflection time levels within a data set to show radar structures at a specified time (depth) across a surveyed site (see e.g. [79, 100]).

5.2.2. Data Interpretation

As previously specified, the GPR raw reflection data stand for a set of individual traces, each containing a series of waves reflected from buried interfaces. For both the 2D and 3D data acquisition techniques, the wave amplitudes, that identify reflection events, are represented as function of two-way travel times (in *ns*). These last can be transformed into depths by assuming a suitable value for the relative dielectric constant that permits to calculate the radiowave velocity from eq. (12). For 3D surveys the depth of the target, or the thickness of a layer, is retrieved by two-way travel time maps that identify reflections over significant areas (see Section 5.2.1). This approach is commonly used for road pavement analysis with the aim to outline the concrete, granular, upper and lower sub-grades layers (see [2, 101]). It is worth to stress that the individuation of the onset of the various reflections on the radargrams as well as the choice of the radiowave velocities for derive the depths require a lot care. Local variability in radiowave velocities, indeed, can occur within concretes due to changes in moisture content that can bring to an uncorrected layer's individuation [2].

Moreover, when a very high resolution is required, as for the non-destructive analysis of edifices or masterpieces, high-level processing and interpretation techniques of GPR data, such as amplitude analysis, reflection coefficient evaluation, definition of ε_r variations, have to be adopted (see i.e. [97, 100]). For example, areas characterised by a high attenuation of the reflected radar energy may reveal high conductivity sectors very likely ascribable to humidity infiltrations within porous or degraded portions of a building as well as concrete structures.

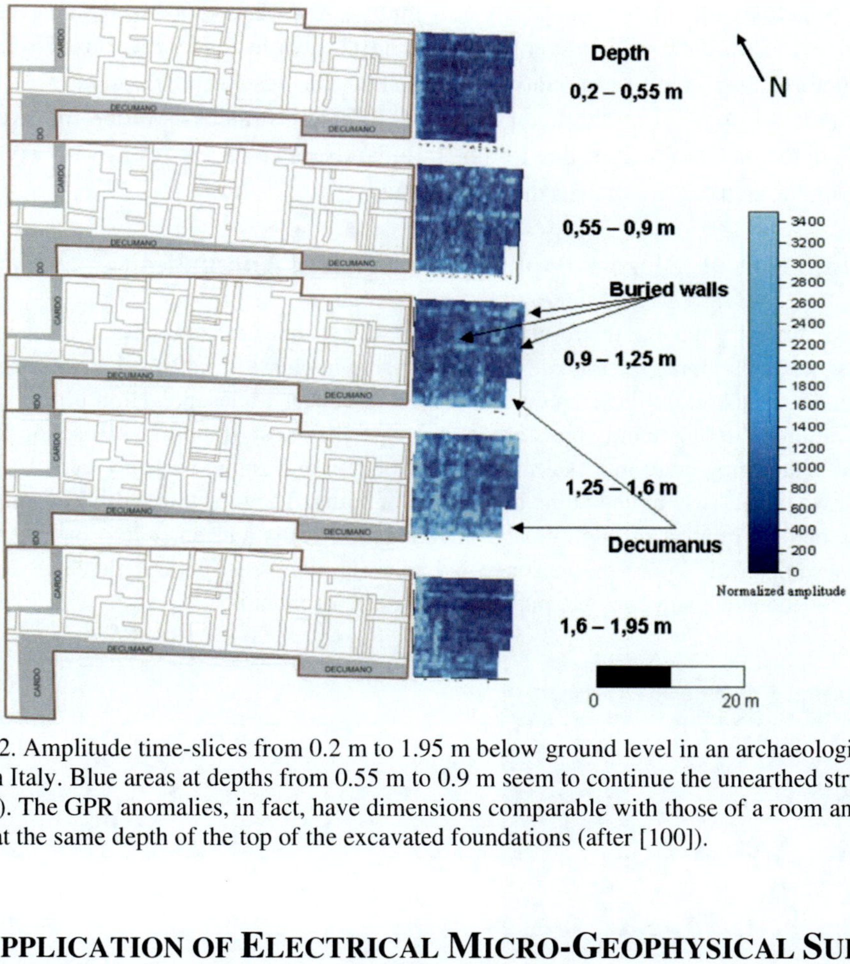

Figure 12. Amplitude time-slices from 0.2 m to 1.95 m below ground level in an archaeological area of Southern Italy. Blue areas at depths from 0.55 m to 0.9 m seem to continue the unearthed structures (on the right). The GPR anomalies, in fact, have dimensions comparable with those of a room and are located at the same depth of the top of the excavated foundations (after [100]).

6. Application of Electrical Micro-Geophysical Surveys for Structural and Physical Characterisation of Architectural Structures

In micro-scale prospecting of architectonic manufactures, it is important to understand the origin of the contrast between the physical properties of the sound and the worsen (o repaired) parts of an artefact. This will permit to define the local structural and physical conditions of the investigated body. For electric and electromagnetic micro-geophysical surveys the principal structural contrasts can be originated by the presence of micro-fracture systems, voids, reinforcement of materials unlike the original. The main physical contrasts, instead, can be due to capillary ascent of humidity and ingression of more or less aggressive waters, which can give to the body particularly conductive internal paths in contrast with resistive situations characterised by compact structures and/or dry pores.

Generally, in the inspection of building's state, the presence of both a high-resistivity anomaly and a meaningless SP anomaly is a good indicator of non-porous or unfractured material, hosted within an otherwise less resistive material with relatively high electric charge distributions, such as more porous and/or fractured blocks permeated by fluids. Air-filled voids would create a similar high-resistivity anomaly and low SP anomaly, but this may be compromised if the void space had suffered a partial or total collapse. Furthermore, the presence of reactive minerals, such as fly ash and/or silica fume, as well as carbonation processes may increase the resistiviy of concrete [102], while the corrosion of reinforced concrete structures may give rise to relatively high SP anomalies (see [103, 104]).

In the next section, some laboratory tests will be illustrated in order to give some guidelines to detect heterogeneities due to moisture, voids, micro-fractures as well as thin and very thin layers in architectural materials and structures.

6.1. Individuation of Micro-Geophysical Electrical Anomalies

As more times pointed, micro-cracks, voids, structural discontinuities, moisture are commonly associated with the material degradation arising in building walls and more generally in any architectural structure; all these factors, in fact, may concur to desegregation of the structure causing its failure. Among them, moisture is perhaps the most important cause of building deterioration. Water may enter into walls as a result of rain and snow, moisture can also rise up within a wall when soil is saturated at the base of the building. The moisture will wet footings and crawl spaces or find its way through cracks in foundation walls and enter into basements, and eventually cause deterioration of the masonry and adjacent wooden structural elements. Moisture is present in new concrete and concrete may hold moisture for months.

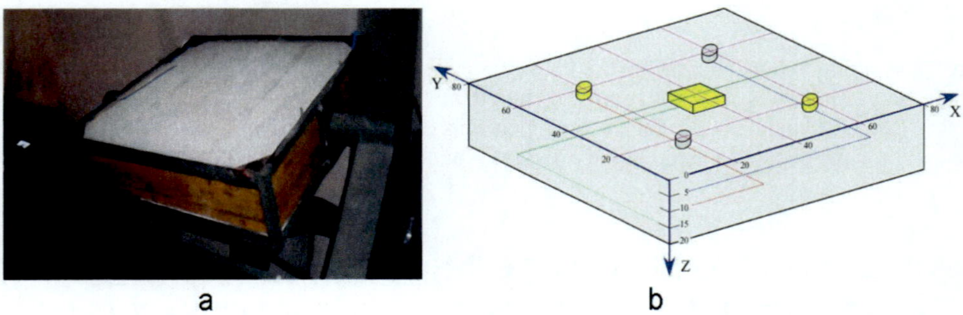

Figure 13. (a) View of the concrete specimen fabricated for the micro-geophysical survey. (b) Sketch of the specimen indicating the position of the defects: yellow and green boxes specify, respectively, tuff and air-filled plexiglas inserts.

In order to study the variations of the electrical parameters characterizing architectural materials in presence of degradation, such as moisture and defects, the results of some laboratory tests, carried out on a concrete specimen, will be shown. The specimen, illustrated in Figure 13, it has a square section with side of 80 cm and thickness of 20 cm. It was fabricated by filling a mould with grout made of a mixture of cement, sand and water. Moreover, some defects were created during fabrication in order to simulate material deterioration. Specifically, two air-filled plexiglas boxes and three tuff pieces were inserted. Plexiglas boxes may simulate cavities which may form if no adequate and uniform pressure is applied during setting; tuff pieces simulate the presence of slag inclusions. Four defects have diameter equal to 7 cm and one defect has a diameter of 10 cm, for all defects the thickness is 3 cm. The survey, consisting on resistivity, self-potential and GPR measurements, was carried out three months after fabrication and repeated again six months later.

6.1.1. Resistivity and SP Measurements

For both resistivity and SP surveys a 3D tomographic technique has been adopted. Taking into account that for most micro-geophysical applications the principal requirements are: *(i)* electrode configurations that allow rapid data acquisition over a densely sampled grid, and *(ii)* non-invasive small-sized electrodes with a low contact resistances at the electrode-concrete interface, a 256–channel instrument and electrodes used in medicine for electrocardiograms (ECG) were used. In particular, the micro-geoelectric System MRS-256 [9], developed by GF Instruments (Czech Republic), is designed for high-resolution resistivity and self-potential measurements to be carried out in the field as well as on architectural structures, such as walls, floors and columns. The system consists of a direct current transmitter (up to 300 V, 1 A) for *AB* electrodes and a multi-channel receiver for simultaneous measurement of current and potential in the grid of electrodes (with input impedance higher than 10 GΩ for all channels and input voltage of ± 5V). The receiver operates using a PC notebook with dedicated software for data acquisition and processing. The system was connected, by means of shielded wires, to the ECG electrodes that allowed to overcome some of the problems (high contact resistance, surface perforation, signal instability, etc.) of conventional electrodes (copper plaquettes, steel needles, etc.). The ECG electrodes, in fact, with a diameter of about 1.5 cm, are made of thin and round metal cap imbedded in a conductive silicon gel which is semi-confined by a small sponge: the gel reduces the contact resistance while the sponge ensures a non-invasive good contact with the structure surface.

The measure grid size was 675 x 675 mm^2 with a mesh of 45 mm in both x and y directions. For the 3D resistivity tomography, sixty-six different current dipoles have been realised along the x direction to produce 8448 measurements, while the SP survey has involved the measure of 480 natural potential drops.

The inversion of the resistivity data has been performed using the Res3Dinv algorithm [31, 32], while the interpretation of the SP measurements has been realised adopting the 3D tomographic inversion method described in the Section 4.2.2.

Figure 14 shows the SP maps obtained three months after fabrication (Figure 14a) and six months later (Figure 14b). By comparing the two maps we may deduce that in the second survey the SP anomaly sources are deeper to respect the first survey due to the presence of relatively low frequency anomalies in the map of Figure 14b. This observation suggests that, six months after specimen fabrication, the moisture is located more in depth with respect to the first survey. The SP data inversion confirms such observations. In fact, as it can be seen in Figure 15 no significant electric charge accumulations are present in depth three months after fabrication (Figure 15a), while six months later (Figure 15b) negative and positive charge accumulation nuclei, located in the area including the defects, are present in the depth range from 2 cm to 12 cm. Moreover, it is also possible to see that on the right side of the specimen is present a zone of negative charge accumulation between two positive nuclei (Figure 15b) which could indicate a variation in concrete characteristics.

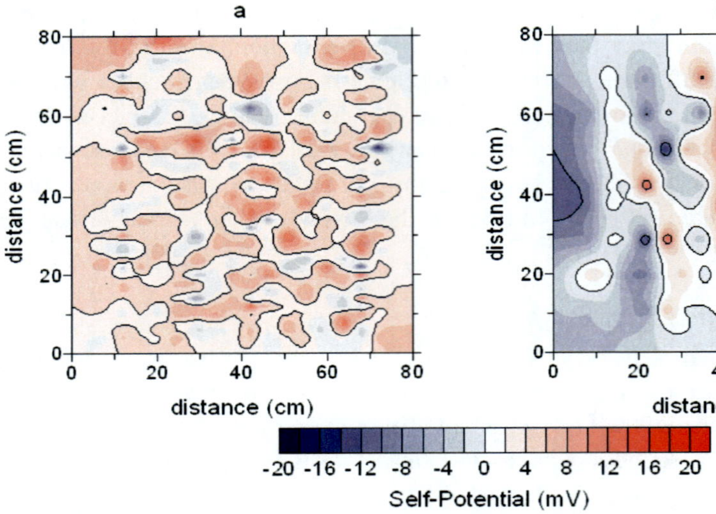

Figure 14. SP maps obtained three months (a) and six months (b) after fabrication of the specimen of Figure 13.

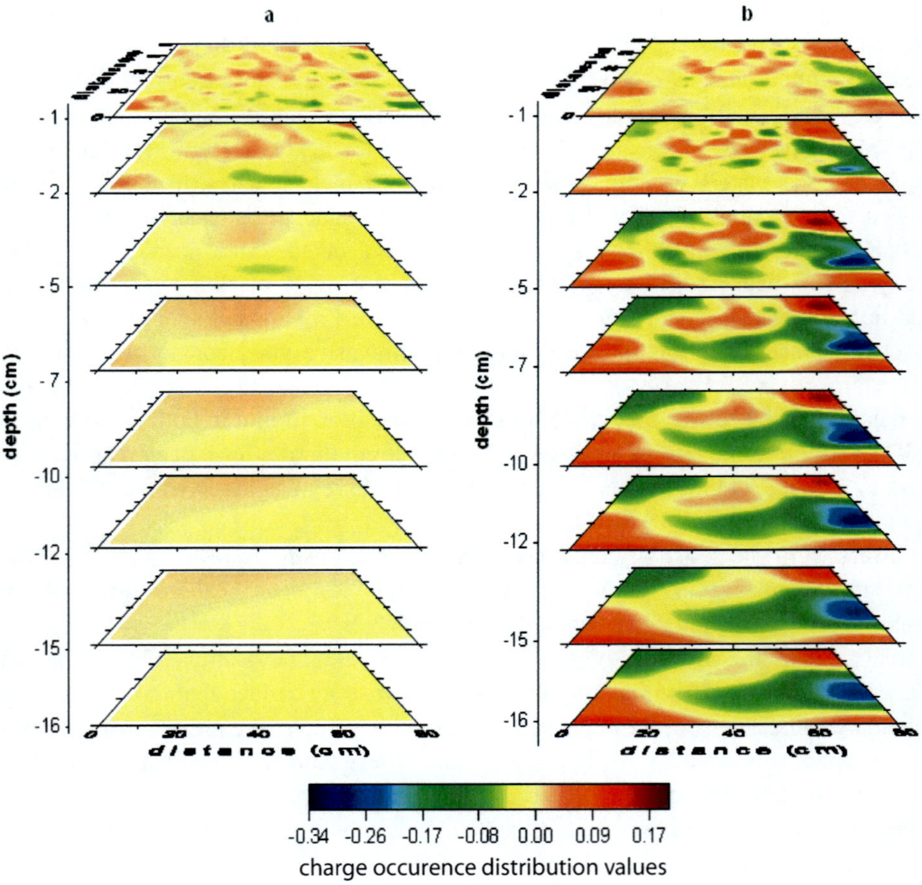

Figure 15. Self-potential 3D tomography images relative, respectively, to the SP data observed three months after specimen fabrication (a) and six months later (b).

Resistivity data in general confirm the self-potential data interpretation. In fact, as outlined by the 3D tomographic representation of the resistivity values observed three months after fabrication (Figure 16a), the specimen is characterized by relatively low resistivity values at depths from 5 cm to 10 cm, which indicate presence of moisure in this depth interval. Six months later (Figure 16b), part of the water evaporated and the remaining fraction occupies deeper portion of the panel.

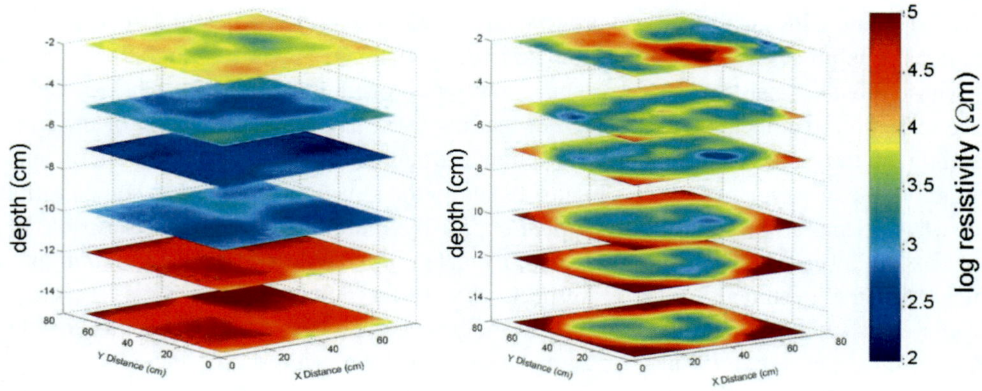

Figure 16. 3D tomographic inversion of the electrical resistivity data observed three months after specimen fabrication (a) and six months later (b).

The resistivity values, in fact, are now higher of about one order of magnitude at medium depths and lower of the same order of magnitude in the central portion of the panel at depths from 12 cm to 15 cm. Moreover, six months after specimen fabrication (Figure 16b), it is also possible to individuate the tuff inserts due to a better contrast between concrete and tuff resistivity. For a better visualization of the defects two data views (Figure 17) are presented through slides parallel to *x-z* plane (Figure 17a) and through slides parallel to *x-y* plane (Figure 17b). Specifically, in the first representation it is possible to see mainly the low depth defects while in the other view the defect at 12 cm (outlined by a black circle) is clearly evident.

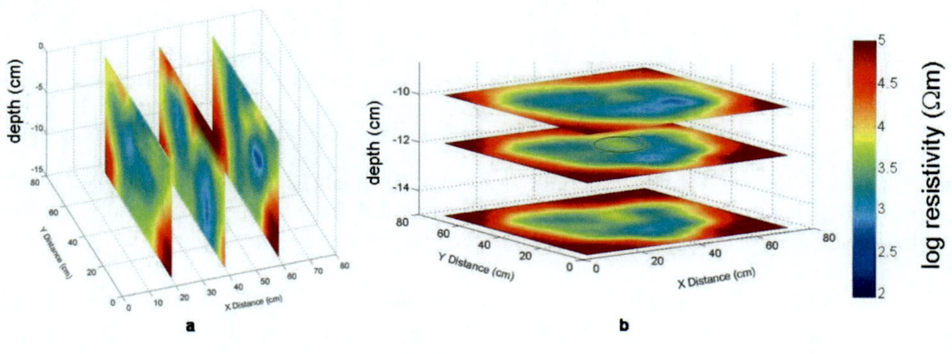

Figure 17. View of the 3D resistivity tomography of Figure 16b through slides parallel to x-z plane (a) and to x-y plane (b).

6.2.2. GPR Measurements

The GPR survey was performed using a GSSI SIR-3000 equipped by a mono-static antenna with centre frequency of 900 MHz. The measurements have been carried out on both the front and the rear face of the specimen of Figure 13 along six profiles of which three parallel to the x-axis and three parallel to the y-axis. The profiles, *20 cm* apart on both the directions, have been located at a distance of 20, 40 and 60 cm from the origin of the reference system (see Figure 13 b). Due to the antenna size, the maximum length of the measurement profiles is found to be *58 cm*, therefore the measure grid size has been of 58 x 58 cm^2 (Figure 18).

The acquired GPR data have been processed to obtain images of the buried defects in terms of radargrams. The processing is consisted on static correction and Kirchoff's migration of the data [96], whose effects consist, respectively, on a translation of the time scale, that brings the upper surface reflection to coincide with the zero of the time-axis, and on an estimation of the size and orientation of the objects causing the reflections through the analysis of the pattern and the amplitude of the observed hyperboles. Figure 19 shows the radargrams relative to some profiles of Figure 18. Looking at the figure, well evident appear the radiowave reflections ascribable to the upper and the lower surfaces of the specimen, indicated in the radargrams along the *hh'* and *aa'* profiles with the symbols *U* and *L*, respectively. These reflections emerge as two regular horizontal bands at about 0 ns and 5 ns, due to the passage between a medium characterised by a relative electric permittivity equal to *1* (air) and a medium with a different relative electric permittivity (specimen), and viceversa. Obviously, the reflections appear flat due to the continuity of the surfaces. The lower surface reflection, in particular, has permitted to calculate the velocity of the block material that is resulted of about 10 cm/ns.

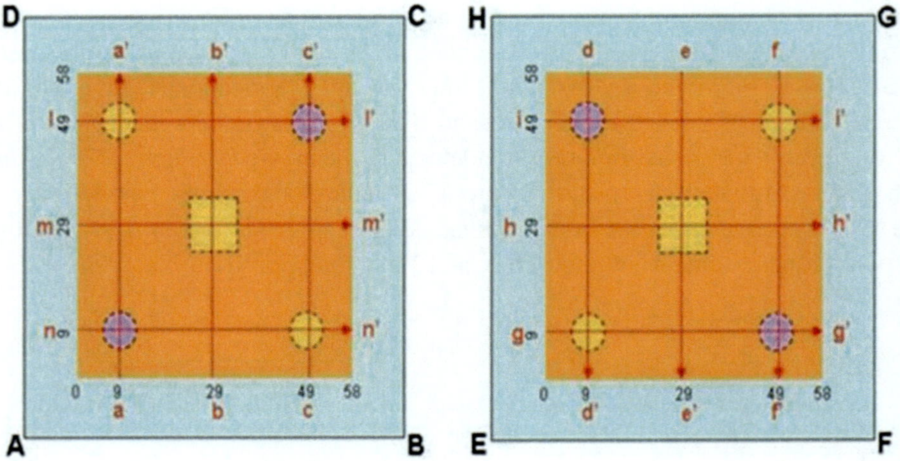

Figure 18. Location of the GPR profiles on the front (ABCD square) and the rear (EFGH square) face of the specimen of Figure 13. The measurement on the rear face have been performed after a rotation of 180° of the specimen around the horizontal axis of the lodging support (see Figure 13a).

Moreover, hyperbolas caused by the inserted defects are also present. Specifically, a rather accentuate hyperbola (*T*) is clearly evident in the radargram along the *hh'* profile, very likely attributable to the central tuff-block (see Figure 13). In fact, the position of the hyperbola along the x-axis and the time at which the reflections are observed well correspond with the position and the distance of the tuff-block from the lower surface (5 cm) of the specimen.

A hyperbola imputable to the same block is visible also in the radargram along the *mm'* profile (Figure 19), but, due to the distance of the tuff-block from the upper surface (12 cm), the reflections here appear much less evident. In fact, the high-frequency antennas are able to provide high resolution data but low penetration depth, hence low electric permittivity contrasts (such as concrete-tuff) are less evident of higher permittivity contrasts (such as concrete-air) as we move away the acquisition surface. This also justifies, as we can see in the radargram along the *mm'* profile, the higher amplitude of the reflection originated by the lower surface (20 cm deep) to respect to that caused by the central tuff-block t_3 (12 cm deep).

As it concerns the hyperbolas indicated in the radargrams of Figure 19 with h_1 and h_2, k_1 and k_2, it is easy to verify that their positions, along the *x*-axis, well correspond to the location of the two air-filled plexiglas boxes and the two small tuff pieces, respectively (see Figure 13). In particular, the hyperbolas h_1 and k_1 shown, respectively, in the *nn'* and *ll'* radargrams are observed at about *1,4* ns, exactly corresponding to the depth (7 cm) of the plexiglass box p_1 and the tuff disk t_1 (see Figure 13). The top of the hyperbolas k_2 and h_2, instead, appears at about 1 ns. Considering that these two hyperbolas represent the reflections due to, respectively, the tuff-disk t_2 and the plexiglass box p_2, located at 2 cm of depth, their top should be noted at 0,4 ns. The discrepancy is very likely ascribable to the proximity of the upper surface of the specimen that could have produced an overlapping of the reflection due to the upper surface with those due to the defects, not permitting their discrimination. Finally, comparing the two radagrams relative to the *ll'* and *nn'* profiles, it is evident that the hyperbolas due to the plexiglass defects (i.e. h_1 and h_2), although located at the same depth of the corresponding tuff defects (i.e. k_1 and k_2), show higher amplitude reflections, outlining a relative electric permittivity contrast between the concrete and the plexiglass greater than the contrast between the tuff and the concrete.

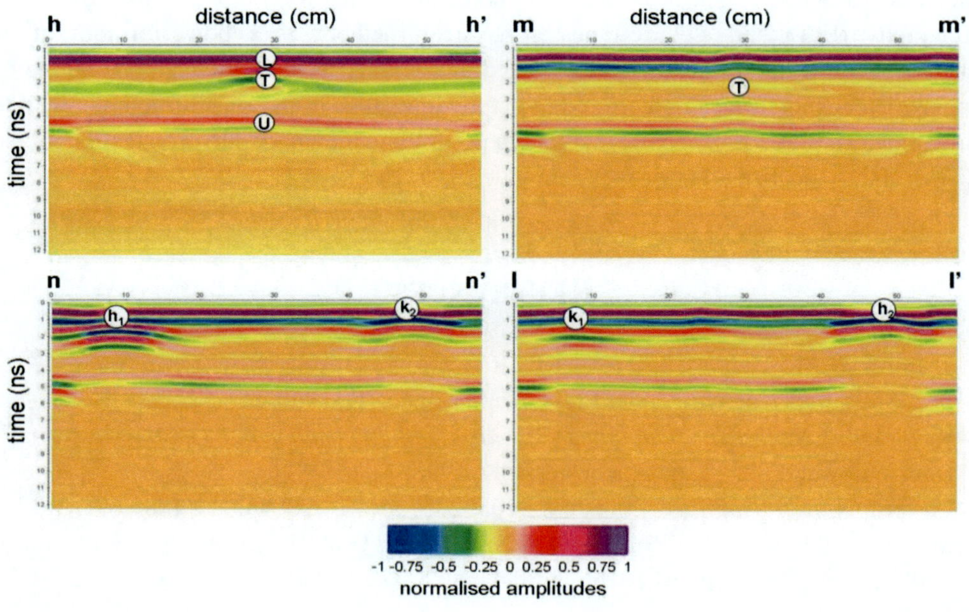

Figure 19. Radargrams relative to the profiles hh', mm', nn' and ll' shown in Figure 18. The hyperbolas: h1, h2 and k1, k2 are very likely produced by the two air-filled plexiglas boxes and the two small tuff pieces, respectively, while the hyperbola T is ascribable to the central tuff-block (see Figure 13). Finally, the two regular horizontal bands at about 0 ns and 5 ns seem to indicate, respectively, the air-specimen and the specimen-air discontinuities.

7. Conclusion

Results of laboratory micro-geophysical measurements on building materials have been illustrated and discussed with the aim to outline the potentiality of the electrical geophysical methodologies in non-destructive control of architectural structures. In particular, the geophysical survey, consisted in the application of the resistivity, self-potential and GPR methods, has pointed out that only a multi-methodological approach is able to outline the heterogeneities ascribable to the main causes of degradation in architectural structures, such as moisture, voids, micro-fractures, occurrence of thin bodies. Specifically, the presence and the time-variations of SP anomalies are indicative, respectively, of underground fluid circulation (i.e. moisture) and of its migration in depth if the wavelength of the anomalies increases with time. Moreover, the distribution of the electric charge concentrations, retrieved from the inversion of the acquired SP data, may help to delineate hidden micro-fracture networks. An estimation of the humidity level, instead, can be provided by resistivity measurements: the higher the observed resistivity values the lower the water content in the investigated material. In addition, resistivity surveys are able to outline small physical and/or structural heterogeneities, such as voids, restored sectors in terms of resistivity contrasts between the homogeneous and the damaged parts of a structure. Finally, the GPR investigation, carried out by using very-high frequency antennas, is suitable to give a detailed image of very-shallow anomalous features in masonry structures, such as the presence of thin and very thin layers of materials unlike the original.

Concluding, it is worth mentioning that results of recent studies [4, 54, 92] devoted to architectural structure inspection have demonstrated that a combined use of infrared thermography (IRT) and electric-type geophysical methods is a powerful tool of non-destructive control. In particular, the IRT technique, jointly to the GPR method, can be used to quickly detect sub-surface defects and moisture anomalies in masonry structures, while the SP and the resistivity prospecting may help to follow the in depth propagation of such anomalies and to localize other deeper anomalies.

References

[1] Telford, W.M.; Geldart, L.P.; Sheriff, R.E.; Keys, D.A. *Applied Geophysics.* 2nd edn; pp. 753, Cambridge University Press: Cambridge, UK, 1990.

[2] Reynolds, J.M. *An Introduction to Applied and Environmental Geophysics*; pp. 796, John Wiley and Sons Ltd: Chichester, UK, 1998.

[3] Linford, N. *Rep Prog Phys* 2006, *69*, 2205–2257, doi:10.1088/0034-4885/69/7/R04.

[4] Meola, C.; Di Maio, R.; Roberti, N.; Carlomagno, G.M. *Engineering Failure Analysis* 2005, *12*, 875-892, doi:10.1016/j.engfailanal.2004.12. 030 (cited in: CSA/ASCE Civil Engineering Abstracts).

[5] Cosentino, P.L. *Proceedings of the 2nd International Conference on Environmental and Engineering Geophysics* June 4-9, 2006, Wuhan, China. Geophysical Solutions for Environment and Engineering – Vol. 2.

[6] Pérez-Gracia, V.; García García, F.; Rodriguez Abad, I. *NDTandE Int* 2008, *41*, 341–353, doi:10.1016/j.ndteint.2008.01.001.

[7] Patella, D.; Piro, S.; Versino, L.; Cammarano, F.; Di Maio, R.; Mauriello, P. *Non-Destructive Techniques Applied to Landscape Archaeology*, Ed.s M.Pasquinucci and F.Trément; The Archaelogy of Mediterranean Landscapes Series (Ed.s G.Barker and D.Mattingly); Oxbow Books: Park End Place, Oxford, 2000, vol.4, 96-103, (ISBN 1 900188 74 0).

[8] Polder, R.B. *Construction and Building Materials* 2001, 15, 125-131.

[9] Cosentino, P.; Di Maio, R.; Meola, C.; Capizzi, P.; Curciotti, R.; Roberti, N.; Soldovieri, M.G. *11th European Meeting of Environmental and Engineering Geophysics, Near Surface 2005*, 4-7 September, 2005 Palermo (Italy), CD-ROM B030.

[10] Ouglova, A.; Raharinaivo, A.; Berthaud, Y.; Petre-Lazar, I.; Boukhenfouf, H. *Materials and Structures* 2005, *38*, 711-719, doi:10.1617/14164.

[11] Newlands, M.D.; Jones, M.R.; Kandasami, S.; Harrison, T.A. *Materials and Structures* 2008, *41*, 621–632, doi:10.1617/s11527-007-9257-6.

[12] Zhdanov, M.S.; Keller, G.V. *The Geoelectrical Methods in Geophysical Exploration*; “Methods in Geochemistry and Geophysics” Series, *31*, pp. 873, Elsevier: Amsterdam, 1998.

[13] Schlumberger, C. *Étude de la prospection électrique du sours-sol* ; Gauthiers-Villars et Cie.: Paris, 1920.

[14] Stefanescu, S.; Schlumberger, C.; Schlumberger., M. *J. de Phys. et du Radium* 1930, *VII*, 132-140.

[15] Tikhonov, A.N. *Dokl. Akad. Nauk SSSR* 1943, *39*, 195-198.

[16] Wenner, F. *U.S. Bureau of Standards Bulletin* 1912, *8*, 559-610.

[17] Broadbent, M.; Habberjam, G.M. *Geophysical Prospecting* 1971, *19*, 321-338.

[18] Barker, R.D. *Geophysical Prospecting* 1981, *29* (1), 128-143.

[19] Worthington, M.H. *First Break* 1984, *2*, 20-26.

[20] van Overmeeren, R.A; Ritsema, I.L. *First Break* 1988, *6*, 313-324.

[21] Dahlin, T. *First Break* 1996, *14*, 275-283.

[22] Loke, M.H.; Barker, R.D. *Geophysical Prospecting* 1996, *44*, 499-523.

[23] Stratton, J.A. *Electromagnetic Theory*; McGraw-Hill: New York, 1941.

[24] Mooney, H.M.; Orellana, E.; Pickett, H.; Tornheim, L. *Geophysics* 1966, *31*, 192–203.

[25] Ghosh, D.P. *Geophysical Prospecting* 1971, *19*, 192-217.

[26] Koefoed, O. *Geosounding Principles, 1 - Resistivity Sounding Measurements*; Elsevier: Amsterdam, 1979.

[27] Constable, S.; Parker, R.; Constable, C. *Geophysics* 1987, *52*, 289-300.

[28] Sasaki, Y. *Geophysics* 1989, *54*, 254-262.

[29] Sasaki, Y. *Geophysical Prospecting* 1992, *40*, 453-464.

[30] Griffiths, D.H.; Barker, R.D. *J Appl Geophys 1993*, *29*, 211-226.

[31] Loke, M.H.; Barker, R.D. *Geophysics* 1995, *60*, 1682-1690.

[32] Loke, M.H.; Barker, R.D. *Geophysical Prospecting* 1996, *44*, 131-152.

[33] Fox, R.W. *Phil Trans R Soc* 1830, *121*, 399-414.

[34] Nyquist, J.E.; Corry, C.E. *The Leading Edge* 2002, *21*, 446-451.

[35] Sato, M.; Mooney, H.M. *Geophysics* 1960, *25*, 226-249.

[36] Ahmad, M. U. *Geophysical Prospecting* 1964, *12*, 49-64.

[37] Keller, G.V.; Frischknecht, F.C. *Electrical methods in geophysical prospecting*; Pergamon Press Inc.: Oxford. 1966.

[38] Zohdy, A.A.R.; Anderson, L.A.; Muffler, L.J.P. *Geophysics* 1973, *38*, 1130-1144.

[39] Zablocki, C.J. (1976). *Proceedings of the 2nd Symposium on the Development and Use of Geothermal Resources, San Francisco*, US Government Printing Office: Washington; Vol. 2., 1299–1309, 1976.

[40] Corwin, R.F.; Hoover, D.B. *Geophysics* 1979, *44*, 226-245.

[41] Ercan, A.; Drahor, M.; Atasoy, E. *Geophysical Prospecting* 1986, *34*, 475-491.

[42] Di Maio, R.; Patella, D. *Acta Vulcanologica* 1994, *4*, 119-124.

[43] Di Maio, R.; Di Sevo, V.; Giammetti, S.; Patella, D.; Piscitelli, S.; Silenziarlo, C. *Annals of Geophysics* 1996, *39*, 179-188.

[44] Corwin, R.F.; Morrison, H.F. *Geophysical Research Letters* 1977, *4*, 171-174.

[45] Fitterman, D.V. *Journal of Geophysical Research* 1978, *83* (B12), 5923-5928.

[46] Varotsos, P. ; Alexopoulos, K. *Tectonophysics* 1984, *110*, 73-98.

[47] Honkura, Y.; Kuro, S. *Journal of Geomagnetism and Geoelectricity* 1986, *38*, 1001-1014.

[48] Di Maio, R.; Patella, D. *Bollettino di Geofisica Teorica ed Applicata* 1991, *XXXIII*, 145-154.

[49] Patella, D.; Tramacere, A.; Di Maio, R. (1997). *Modelling earth current precursors in earthquake prediction*, Ed.s A.Vogel and K.Brandes (Ed.s); «Mitigating the Impact of Impending Earthquakes», A.A.Balkema: Rotterdam, 1997, 121-146, (ISBN 90 5410 801 0).

[50] Wynn, J.C., Sherwood, S.I. (1984). *The self-potential (SP) method: an inexpensive reconnaissance and archaeological mapping tool*. Journal of Field Archaeology, 11, 195-204.

[51] Drahor, M.G., Akyol, A.L,, Dilaver, N. (1996). *An application of the self-potential (SP) method in archaeogeophysical prospection*. Archaeological Prospection, 3, 141-158.

[52] Di Maio, R.; Fedi, M.; Garofalo, B.; La Manna, M.; Roberti, N.; Soldovieri, M.G. *Rivista di Studi Pompeiani* 2009, *XIX*, 115-123.

[53] Drahor, M.G. *Archaeological Prospection* 2004, *11*, 77-105.

[54] Carlomagno, G.M.; Di Maio, R.; Meola, C.; Roberti, N. *QIRT Journal* 2005, *2*, 5-24.

[55] Corwin, R.F. *The self-potential method for environmental and engineering applications*. In *Geotechnical and Environmental Geophysics*, v. 1, Ed. Ward S.H., Tulsa, Oklahoma: Society of Exploration Geophysicists, 1990, 127-145.

[56] Rao, B.S.R.; Murthy, I.V.R.; Reddy, S.J. *Pure and Applied Geophysics* 1970, *78*, 66-77.

[57] Fitterman, D.V. *Geophysics* 1979, *44*, 195-205.

[58] Bhattacharya, B.B.; Roy, N. *Geophysical Prospecting* 1981, *29*, 102-107.

[59] Murty, B.V.S.; Haricharan, P. *Geophysics* 1985, *50*, 1127-1135.

[60] Fitterman, D.V.; Corwin, R.F. *Geophysics* 1982, *47*, 938-945.

[61] Ram Babu, H.V.; Atchuta, R.D. *Computer and Geosciences* 1988, *14*, 377-387.

[62] Jagannadha, R.S.; Rama, R.P.; Radhakrishna, M. *Computer and Geosciences* 1993, *19*, 61-73.

[63] Marquardt, D.W. *Journal of the Society of Industrial and Applied Mathematics* 1963, *11*, 431-441.

[64] Patella, D. *Geophysical Prospecting* 1997, *45*, 653-681.

[65] Leimbach, G.; Löwy, H. *German patent* n. 237944, 1910.

[66] Hülsenbeck and Co. *German patent* n. 484934, 1926.

[67] Bogorodsky, V.V.; Bentley, C.R.; Gudmansen, P.E. *Radioglaciology*;: D. Reidel Publishing Co.: Dordrecht, 1985.
[68] Unterberger, R.R. *Geophysical Prospecting* 1978, *26* (2), 312-328.
[69] Davis, J.L.; Annan, A.P. *Geophysical Prospecting* 1989, *37* (5), 531-551.
[70] Moorman, B.J.; Judge, A.S.; Smith, D.G. *Examining fluvial seediments using ground penetrating radar in British Columbia*; Current Research: Part A. Geological Survey of Canada 1991, 91-1A, 31-36.
[71] Huggenberger, P.; Meier, E.; Pugin, A. *J Applied Geophysics* 1994, *31* (1-4), 171-184.
[72] Neal, A.; Pontee, N.I.; Pye, K.; Richards, J. *Sedimentology* 2002, *49* (4), 789-804.
[73] Leggo, P.J.; Leech, C. *Ground Engineering* 1983, *16*, 20-23.
[74] Darracot, B.W.; McCann, D.M. *Planning engineering geophysical survey*, 1986. In: Hawkins, A.B. (ed.), Site Investigation Practice: Assessing BS 5930, Geological Society Engineering Geology Special Publication no.2, 85-90.
[75] Clemena, G.G.; Sprinkel, M.M.; Long Jr, R.R. *Transp. Res. Rec.* 1987, *1109*, 1-10.
[76] Saarenketo, T.; Scullion, T. *J Applied Geophysics* 2000, *43* (1-3), 119-138.
[77] Bevan, B.W.; Kenyon, J. *MASCA Newsletter* 1975, *11* (2), 2-7.
[78] Imai, T.; Sakayama, T.; Kanemori, T. *Geophysics* 1987, *52* (2), 137-150.
[79] Malagodi, S.; Orlando, L.; Piro, S.; Rosso, F. *Archaeological Prospection* 1996, *3*, 13-23.
[80] Tomizawa, Y.; Arai, I.; Hirose, M.; Suzuki, T.; Ohhashi, T. *Archaeological Prospection* 2000, *7*, 241-247.
[81] Barone, P.M.; Graziano, F.; Pettinelli, E.; Ginanni Corradini, R. *Archaeological Prospection* 2007, *14* (1), 47-59.
[82] Vannaroni, G.; Pettinelli, E.; Ottonello, C.; Cereti, A.; Della Monica, G.; Del Vento, D.; Di Lellis, A.M.; Di Maio, R.; Filippini, R.; Galli, A.; Meneghini, A.; Orosei, R.; Orsini, S.; Pagnan, S.; Paolucci, F.; Pisani, A.R.; Schettini, G.; Storini, M.; Sacconi, G. *Planetary and Space Science* 2004, *52*, 67-78.
[83] Pettinelli, E.; Burghignoli, P.; Galli, A.; Pisani, A.R.; Ticconi, F.; Vannaroni, G.; Bella, F. *IEEE-TGRS* 2007, *48*, 1271-1281.
[84] Al-Qadi, I.L.; Lahouar, S. *Constr Build. Mater* 2005, *19*, 763-772.
[85] Loizos, A.; Plati, C. *NDTandE Int* 2007, *40*, 147-157.
[86] Shaari, A.; Millard S.G.; Bungey, J.H. *NDTandE Int* 2004, *37*, 237-242.
[87] Lorenzo, H.; Cuéllar, V.; Hernández, M.C. *J Applied Geophysics* 2001, *47*, 327-336.
[88] Grandjean, G.; Gourry, J.C.; Bitri, A. *J Applied Geophysics* 2000, *45*, 141-156.
[89] Porsani, J.L.; Sauck, W.A.; Júnior, A.O.S. *J Applied Geophysics* 2006, *58*, 177-187.
[90] Muldoon, R.; Chalker, A.; Forde, M.C.; Ohtsu, M.; Kunisue, F. *Constr Build Mater* 2007, *21*, 527-537.
[91] Grote, K.; Hubbard, S.; Harvey, J.; Rubin, Y. *J Applied Geophysics* 2005, *57*, 129-153.
[92] Garofalo, B.; La Manna, M.; Meola, C.; Soldovieri, M.G.; Tiano, P.; Di Maio, R.; Fedi, M.; Roberti, N. *XXVI Congress GNGTS-CNR*, Rome (IT), 13-15 november 2007, Extended Abstracts, 476-477.
[93] Bungey, J.H. *Constr Build Mater* 2004, *18*, 1-8.
[94] Barrile, V.; Pucinotti, R. *NDTandE Int* 2005, *8*, 596-604.
[95] Shaw, M.R.; Millard, S.G.; Molyneaux, T.C.K.; Taylor, M.J.; Bungey, J.H. *NDTandE Int* 2005, *38*, 203-212.

[96] Annan, A. P. *Ground Penetrating Radar Principles, Procedures and Applications*; pp. 293, Mississauga, ON, Canada, Sensors and Software, 2004.

[97] Conyers, L.; Goodman, D. *Ground Penetrating Radar: An Introduction for Archaeologists*; pp. 232, AltaMira Press: U.S., 1997.

[98] Cosentino, P.; Martorana, R.; Miraglia, D.; Sanfratello, V.; Terranova, L.M. *Proceedings of the 6th Meeting of EEGS* 2000, European Section, Bochum, CH07.

[99] Leckebusch, J. *Archaeological Prospection* 2000, *7*, 189-200.

[100] Fedi, M.; Florio, G.; Garofalo, B.; La Manna, M.; Pellegrino, C.; Rossi, A.; Soldovieri, M.G. *Annals of Geophysics* 2008, *51* (5/6), 867-875.

[101] Saarenketo, T.; Scullion, T. *J Applied Geophysics* 2000, *43*, 119-138.

[102] Polder, R.; Andrade, C.; Elsener, B.; Vennesland, O.; Gulikers, J.; Weidert, R.; Raupach, M. *Materials and Structures* 2000, *33*, 603-611.

[103] Elsener, B.; Andrade, C.; Gulikers, J.; Polder, R.; Raupach, M. *Materials and Structures* 2003, *36*, 461-471.

[104] Ouglova, A.; Raharinaivo, A.; Berthaud, Y.; Petre-Lazar, I.; Boukhenfouf, H. *Materials and Structures* 2005, *38*, 711-719.

INDEX

C

D

E

F

G

H

I

J

K

L

M

S